BUILDING AN EFFECTIVE ENVIRONMENTAL MANAGEMENT SCIENCE PROGRAM:

FINAL ASSESSMENT

AF215421

Committee on Building an Environmental Management Science Program

Virtual Commission on Environmental Management Science

National Research Council

NATIONAL ACADEMY PRESS

Washington, D.C. 1997

NOTICE: The project that is the subject of this report was approved by the Governing Board of the National Research Council, whose members are drawn from the councils of the National Academy of Sciences, the National Academy of Engineering, and the Institute of Medicine. The members of the committee responsible for the report were chosen for their special competences and with regard for appropriate balance.

This report has been reviewed by a group other than the authors according to procedures approved by a Report Review Committee consisting of members of the National Academy of Sciences, the National Academy of Engineering, and the Institute of Medicine.

This work was sponsored by the U.S. Department of Energy, Contract No. DE-FC01-94EW54069/R. All opinions, findings, conclusions, and recommendations expressed herein are those of the authors and do not necessarily reflect the views of the Department of Energy.

Library of Congress Catalog Card Number 97-65687
International Standard Book Number 0-309-05730-2

Additional copies of this report are available from:

National Academy Press
2101 Constitution Ave., NW
Box 285
Washington, DC 20055
800-624-6242
202-334-3313 (in the Washington Metropolitan Area)
http://www.nap.edu

Copyright 1997 by the National Academy of Sciences. All rights reserved.

Printed in the United States of America

COMMITTEE ON BUILDING AN ENVIRONMENTAL MANAGEMENT SCIENCE PROGRAM

JOHN F. AHEARNE, *Chair,* Sigma Xi, and Duke University, Research Triangle Park, North Carolina
EDWARD M. ARNETT, Duke University, Durham, North Carolina
STANLEY I. AUERBACH, SENES Oak Ridge, Inc., Oak Ridge, Tennessee
EDWARD J. BOUWER, Johns Hopkins University, Baltimore, Maryland
JOHN I. BRAUMAN, Stanford University, Stanford, California
NAOMI H. HARLEY, New York University Medical Center, New York
HAROLD LEWIS, University of California (retired), Santa Barbara
DEREK R. LOVLEY,[*] University of Massachusetts, Amherst
ALEXANDER MacLACHLAN, DuPont (retired), Wilmington, Delaware
GENE G. MANNELLA, Gas Research Institute (retired), Potomac, Maryland
NORINE E. NOONAN, Florida Institute of Technology, Melbourne
JEROME SACKS, National Institute of Statistical Sciences, Research Triangle Park, North Carolina
ALFRED P. SATTELBERGER, Los Alamos National Laboratory, New Mexico
LEON T. SILVER, California Institute of Technology, Pasadena

Consultants

GREGORY R. CHOPPIN, Florida State University, Tallahassee
DONALD J. DePAOLO, University of California, Berkeley
GEORGE M. HORNBERGER, University of Virginia, Charlottesville

Staff

KEVIN D. CROWLEY, *Study Director*[†,‡]
TAMAE MAEDA WONG, *Senior Staff Officer*[§]
SUSAN B. MOCKLER, *Research Associate*
ERIKA L. WILLIAMS, *Research Assistant*
PATRICIA A. JONES, *Senior Project Assistant*
DENNIS L. DuPREE, *Senior Project Assistant*

---

[*]Resigned from committee on October 17, 1996.
[†]Board on Radioactive Waste Management.
[‡]Board on Earth Sciences and Resources (through May 1996).
[§]Board on Chemical Sciences and Technology.

iii

VIRTUAL COMMISSION ON ENVIRONMENTAL MANAGEMENT SCIENCE

PERRY L. McCARTY, *Chair,* Stanford University, Stanford, California
RICHARD A. CONWAY, Union Carbide Corporation, South Charleston, West Virginia
DONALD J. DePAOLO, University of California, Berkeley
DAVID J. GALAS, Darwin Molecular Corporation, Bothell, Washington
MICHAEL C. KAVANAUGH, Malcolm Pirnie, Oakland, California
ROYCE W. MURRAY, University of North Carolina, Chapel Hill

Staff

STEPHEN RATTIEN, *Executive Director*
GREGORY SYMMES, *Reports Officer*
JAMES MALLORY, *Administrative Officer*
SANDI FITZPATRICK, *Administrative Associate*

The Committee on Building an Environmental Management Science Program is a joint activity of the Commission on Engineering and Technical Systems; Commission on Geosciences, Environment, and Resources; Commission on Life Sciences, and Commission on Physical Sciences, Mathematics, and Applications.

The National Academy of Sciences is a private, nonprofit, self-perpetuating society of distinguished scholars engaged in scientific and engineering research, dedicated to the furtherance of science and technology and to their use for the general welfare. Upon the authority of the charter granted to it by the Congress in 1863, the Academy has a mandate that requires it to advise the federal government on scientific and technical matters. Dr. Bruce Alberts is president of the National Academy of Sciences.

The National Academy of Engineering was established in 1964, under the charter of the National Academy of Sciences, as a parallel organization of outstanding engineers. It is autonomous in its administration and in the selection of its members, sharing with the National Academy of Sciences the responsibility for advising the federal government. The National Academy of Engineering also sponsors engineering programs aimed at meeting national needs, encourages education and research, and recognizes the superior achievements of engineers. Dr. William A. Wulf is interim president of the National Academy of Engineering.

The Institute of Medicine was established in 1970 by the National Academy of Sciences to secure the services of eminent members of appropriate professions in the examination of policy matters pertaining to the health of the public. The Institute acts under the responsibility given to the National Academy of Sciences by its congressional charter to be an adviser to the federal government, and, upon its own initiative, to identify issues of medical care, research, and education. Dr. Kenneth Shine is president of the Institute of Medicine.

The National Research Council was organized by the National Academy of Sciences in 1916 to associate the broad community of science and technology with the Academy's purposes of furthering knowledge and advising the federal government. Functioning in accordance with general policies determined by the Academy, the Council has become the principal operating agency of both the National Academy of Sciences and the National Academy of Engineering in providing services to the government, the public, and the scientific and engineering communities. The Council is administered jointly by both Academies and the Institute of Medicine. Dr. Bruce M. Alberts and Dr. William A. Wulf are chairman and interim vice-chairman, respectively, of the National Research Council.

PREFACE

The Committee on Building an Environmental Management Science Program was established under the auspices of the National Research Council at the request of Thomas P. Grumbly, Under Secretary of Energy, to advise the Department of Energy on the structure and management of the Environmental Management Science Program (EMSP)—a mission-directed basic research program to support cleanup of the nation's nuclear weapons complex. The committee met seven times from May to November 1996 and produced three reports: an initial assessment report[1] that addresses the near-term needs of the program related to the fiscal year 1996 proposal competition; a letter report[2] that addresses the development of a fiscal year 1997 program announcement; and the present report, which addresses longer-term challenges and opportunities for the program. The statement of task for this report is given in Appendix A under Activity #2: Science and Management Needs.

The DOE cleanup program is the federal government's largest environmental program. The length of time estimated to complete the cleanup task and the dollars estimated to be spent make this program the largest environmental program of any nation. But the program faces many problems that will require new knowledge and fundamental understanding of basic chemical, physical, geological, and biological processes and their relationship to risk. The EMSP, a small and new program, has as its goal to develop that basic knowledge, and this report and its predecessors have the goal of assisting the Department in structuring and managing the EMSP.

The production of three reports in an 8-month period was an extremely difficult task and could not have been accomplished without a dedicated committee and staff. The committee's first meeting was held on Mother's Day weekend, and the second meeting was held on Father's Day weekend. By the third meeting, a semblance of sanity had settled on

---

[1]National Research Council. 1996. Building an Effective Environmental Management Science Program: Initial Assessment. Washington, D.C.: National Academy Press. This report is reprinted in Appendix F and is available on the World Wide Web at the following address: http://www.nap.edu/readingroom/books/envmanage/index.html.

[2]Letter Report to the Associate Deputy Assistant Secretary for Science and Risk Policy, October 8, 1996 (Appendix G).

the committee, which was able to schedule all but one of its remaining meetings during "normal" working hours.

Despite the large task to be accomplished in a short time, the committee reached a near consensus on all issues. Dr. Hal Lewis has included a supplementary statement in Appendix D noting his disagreement with the committee on a few of its conclusions. I have responded to Dr. Lewis's concerns in Appendix E.

I wish to extend my personal thanks to the committee—especially its vice-chair, Norine Noonan—and the committee's three consultants for their diligent work on this project. On behalf of the committee, I also wish to thank the DOE headquarters staff, national laboratory staff, DOE contractor staff, and the many other individuals (see Appendix B) who provided information for this study and answered the committee's many questions. The committee particularly wishes to acknowledge the efforts of Carol Henry, Mark Gilbertson, and Steve Domotor from the Office of Environmental Management; Michelle Broido, Ari Patrinos, and Roland Hirsch from the Office of Energy Research; and Terry Surles and Sally Benson from the Strategic Laboratory Council.

Finally, the committee wishes to thank the staff of the National Research Council for their help with this study: Tamae Maeda Wong for help with meeting organization and report writing, Erika Williams and Susan Mockler for report research, and Tricia Jones and Dennis DuPree for meeting and committee support. This report reflects the great effort, considerable insight, and writing skills of the Study Director Kevin Crowley.

John F. Ahearne, Chair

CONTENTS

BUILDING AN EFFECTIVE ENVIRONMENTAL MANAGEMENT SCIENCE PROGRAM:

FINAL ASSESSMENT

SUMMARY

The Department of Energy's (DOE's) Environmental Management Science Program (EMSP) was created by the 104th Congress to stimulate basic research and technology development for cleanup of the nation's nuclear weapons complex. The EMSP is a mission-directed basic research program and is designed to support a much larger technology development program within the Office of Environmental Management (EM). The program is managed jointly by EM and the Office of Energy Research (ER). Unlike other federal programs that address environmental problems, the EMSP is explicitly focused on EM's problems and has the specific objective to improve the effectiveness of the cleanup effort over the long term.

This is the third of three reports written by this committee at the request of Thomas P. Grumbly, Under Secretary of Energy, to provide advice to the Department on the structure and management of the EMSP.[1] Summaries of the committee's principal conclusions and recommendations are provided in the following sections. More detailed explanations and supporting discussions can be found in the text of the report.

VALUE OF EMSP TO THE DOE CLEANUP MISSION

Many of EM's cleanup problems cannot be solved or even managed efficiently and safely with current technologies, in part owing to their tremendous size and scope. However, cleanup would benefit greatly from the involvement of basic researchers, as noted in recent NRC and DOE reports (see Chapter 2). The committee believes that a basic research program focused on EM's most difficult cleanup problems may have a significant long-term impact on the EM mission. Basic research can provide new knowledge to allow the Department to attack cleanup

---

[1]The other two reports completed during this study are (1) National Research Council, 1996, Building an Effective Environmental Management Science Program: Initial Assessment (Washington, D.C.: National Academy Press), and (2) Letter Report to the Associate Deputy Assistant Secretary for Science and Risk Policy, October 8, 1996. These reports are discussed in Chapter 1 and are reproduced in Appendixes F and G.

problems that are currently intractable or exorbitantly expensive using current technologies; it can lead to the development of better technologies to allow current cleanup to be accomplished at lower costs or with fewer hazards to workers and the public; it can improve understanding of risks and how to discuss them with local stakeholders; and it can lead to the development of new or improved technologies that will allow cleanup to a higher state than is presently possible, thereby making sites available for less restrictive uses. Simply put, new technologies are required to deal with EM's most difficult problems, and new technologies demand new science.

The EMSP is different in several respects from other federal basic research programs, including other DOE programs, that support fundamental research related to the environment. Although several federal programs support basic research in fields broadly relevant to environment science, none are focused explicitly on EM's problems, and none have an explicit link to the problem holders at the sites. In addition, the EMSP will promote the development of partnerships among universities, national laboratories, other federal agencies, and the private sector. These partnerships can bring together highly creative and innovative researchers, provide access to unique national research facilities, and provide a multidisciplinary focus on EM's most difficult problems.

Funding for the EMSP should be viewed as an investment that may, in the long term, lead to more effective cleanup. The EMSP alone will not solve all of EM's cleanup problems—but given the sheer magnitude of the cleanup mission and its estimated cost, coupled with the technological challenges, the committee views the investment in EMSP as both prudent and timely.

DEVELOPMENT OF AN EMSP SCIENCE PLAN

If the EMSP is to have a significant impact on the cleanup mission, the Department must incorporate this program into its strategic plans. Indeed, as the deadline for the Government Performance and Results Act's reporting requirements draws near, it is essential to the survival of the EMSP that a plan for applying basic research in the cleanup program—a *science plan*—be explicitly and officially articulated by the Department. **The committee recommends that the Department** develop a science plan for the EMSP. **This science plan should provide a comprehensive list of significant cleanup problems in the nation's nuclear weapons complex that can be addressed through basic research and a strategy for addressing them.** This science plan should serve as the primary guiding document for the Department's research investment in the cleanup mission.

The committee recommends both a near-term and a long-term process for developing a science plan for the EMSP. For the near term (i.e., the fiscal year 1997 [FY97] competition), the committee recommends that the Department develop a science plan from existing Department documents. Examples of documents that could be used for this purpose are provided in Chapter 3. **For the longer term (i.e., the FY98 competition), the committee recommends that the Department consult with its "problem holders"—the technical staff, managers, and stakeholder advisory groups at the sites who have some understanding of cleanup issues—to obtain guidance on cleanup problems that cannot be addressed practically or efficiently with current knowledge or technologies.** The committee recognizes, of course, that the technical expertise and knowledge for assessing cleanup problems among these groups is uneven and, consequently, suggestions from these groups will have to be considered against that knowledge. Given the large number of DOE sites, these consultations will have to be structured carefully to be manageable by and useful to EMSP staff.

The committee's Letter Report encouraged the Department to broaden its research solicitations and to include problems related to risk, health assessment, and quantitative methodologies (i.e., statistical methods, numerical [simulation] methods and the combination of the two sets of techniques), mainly because the committee believes that research in these areas could have a direct impact on the cleanup mission. In addition, the committee believes that ER should ensure that the pertinent merit review panelists are knowledgeable in the risk research field.

COORDINATING THE INVESTMENT IN BASIC RESEARCH

The science plan is likely to be very broad in scope—both in terms of the range of problems and the disciplinary coverage—and will likely require an investment in basic research that is larger than the current $50 million annual investment in the EMSP. To implement the science plan, Department staff should find ways to utilize relevant research being sponsored in other federal programs and to focus the EMSP on those problems that are unique to the weapons complex.

Given the relatively small size of the EMSP and its staff, the committee does not deem it prudent to recommend formal coordination mechanisms between the EMSP and other research programs. The committee does, however, offer several examples of the kinds of coordinating activities that could be of value to the program in Chapter 3.

BROADENING THE INVESTIGATOR COMMUNITY

Department staff should strive to broaden the community of investigators involved in the EMSP and to expand the core or "committed cadre" of investigators who are knowledgeable about EM's problems. The Department can broaden the community of investigators concerned with its cleanup problems by encouraging (but not requiring) appropriate collaborations among university, industry, and national laboratory researchers. These collaborations are not an end in themselves but rather a route for stimulating new research, introducing new investigators to the Department's problems, and assuring relevance of the projects. By additional encouragement of graduate and postdoctoral training in areas of interest, the Department can further broaden the community of investigators over the longer term.

The committee recommends that collaborations be encouraged where appropriate—but they should not be a requirement for the program. The committee also reaffirms the recommendation from its Letter Report (p. 4) that the program "should encourage (but not require) graduate student involvement in research proposals submitted to the program." The committee would add to this recommendation that appropriate postdoctoral training opportunities, including training opportunities within current DOE programs, also should be encouraged to sustain the interest of talented young scientists.

PROPOSAL SELECTION PROCESS

Based on its review of the data received from the Department, the committee reached the following conclusions about the proposals selected for funding in the FY96 competition: (1) meritorious projects appear to have been selected; (2) collaborative efforts were well represented among the list of successful projects; (3) the program appears to have been successful in attracting some "new" (to DOE) researchers to the program; and (4) in the one case where firsthand information was available, the committee was able to confirm the overall quality of the merit review panel.

The committee has two concerns about the transparency and technical credibility of the merit review process used in the FY96 competition. First, the merit review process was "opaque" to those who submitted proposals to the program and the broader research community. Second, the merit review panels were not allowed to reach consensus on individual proposals or to provide ER program managers with a ranking of proposals because the panels were not constituted under the Federal Advisory Committee Act (FACA). **The committee recommends that the Department examine the entire review process for the EMSP with the goal of increasing its transparency and technical credibility. To this end, the committee recommends that the Department carry through on its stated intention (in its response to a 1991 General Accounting Office report) to seek a change in its legislation to allow FACA proposal review panels—and to convene the EMSP merit review panels under FACA once this change is made.**

The committee also is concerned about the lack of timely feedback to proposers in the FY96 proposal competition. In at least some instances, panelist reviews were not sent to principal investigators (P.I.s) unless requested, and these reviews did not always reflect the discussions in the panel meetings. **The committee recommends that in future competitions the proposal reviews be modified to reflect the discussions at the panel meetings and, further, that applicants receive feedback on the content and result of the reviews in a timely fashion.**

PROGRAM FUNDING

The committee remains concerned about the developing "mortgage" on future-year budgets in the program from commitments made in the FY96 proposal competition. Based on its analysis of future-year funding (Chapter 4), the committee reached the following conclusions about the budget for the program: (1) the annual budget for the EMSP will have to increase significantly to maintain a reasonable number of new starts with an equitable distribution of funding between DOE and non-DOE performers or (2) if the budget remains at current levels, both non-DOE and DOE performers could see about a 75 percent drop in funding for new and competitive renewal projects. The committee believes that, without some assurance that funding will be available to support a reasonable number of new awards annually, EMSP will simply not be viewed as "worth the effort" by potential proposers.

The committee appreciates the difficult budget environment that DOE now finds itself in and recognizes that any increases in the budget for the EMSP may be at the expense of other Department programs. In the committee's view, however, this funding should not come from existing ER programs, which are vital to the Department's long-term mission and are an important part of the nation's basic research portfolio. Nevertheless, the EMSP cannot live up to its potential without careful consideration by DOE of both the total funding levels and the funding patterns (i.e., the balance between new and continuing awards). **The committee urges DOE to find a solution to the problem of not being able to "forward fund" projects at national laboratories and reiterates its recommendation from the previous reports to fully fund all awards in the first year.**

ROLE OF "STAKEHOLDERS" IN
PROPOSAL REVIEW AND SELECTION

The committee does not believe that stakeholders should be involved in the day-to-day management of the program and, in particular, the proposal review and selection process. To be effective and credible, the review and selection process should be carried out by technical experts and should remain free of local concerns and special-interest pressures. Stakeholders should be consulted for guidance on site problems for the EMSP science plan. The committee suggests a process in Chapter 3 for obtaining this guidance.

The committee also believes that participation of EMSP investigators in the proposal selection process would be very helpful in future years. These individuals can bring an important perspective that helps link EMSP more closely to the broad research community, which will benefit the process of shaping the longer-term character of the program.

DOE should also improve and enhance the ways in which it informs the potential users of EMSP results (e.g., technology managers at the various sites) about the process and the outcome of EMSP proposal selection. The hoped-for result of such improved information flow is that these "problem holders" will become more attuned to the long-term benefits of EMSP to their efforts.

LONG-TERM MANAGEMENT STRATEGIES

The committee believes that simplification of program management and a clearer delineation of responsibilities among all management participants is needed to ensure the long-term effectiveness of the EMSP. **To this end, the committee recommends that management of the EMSP be vested in a single individual—an EMSP Program Director—who should have authority, responsibility, and accountability for meeting the program's objectives.**

This Program Director must be involved in the planning activities of both EM and ER and must have the support of the Director of Energy Research and the Assistant Secretary for Environmental Management to utilize the considerable resources from both organizations for the benefit of the EMSP. At the same time, the Program Director must be able to balance the interests of ER and EM and must have the independence to resolve conflicts when these interests come into competition. **To allow for such independence, the committee recommends that the EMSP Program Director report to the Under Secretary for Energy.**

The committee recognizes that this recommendation might be seen by some in the Department as unrealistic when the small size of this program is considered against the other responsibilities of the Under Secretary. Nevertheless, the committee makes this recommendation because it believes that, although the program is small, the success of the

EMSP can be vital to the Department's ability to resolve the contamination legacy and to utilize effectively the several hundred billion dollars estimated to be spent on the cleanup effort.

MAINTAINING PROGRAM QUALITY

To maintain the quality of the EMSP, the committee recommends that the Department convene an independent review panel at appropriate intervals to review the performance and effectiveness of the following aspects of the program:

- merit and relevance review processes,
- quality of funded proposals,
- effectiveness of the application of research results to technology development and cleanup,
- effectiveness of the program in attracting outstanding researchers and innovative research ideas, and
- overall management efficiency and effectiveness.

ASSESSING OUTCOMES

The Department must provide information about performance of the EMSP to meet the requirements of the Government Performance and Results Act of 1993. The committee believes that the best way to assess the performance of the EMSP is through independent peer review. Such review will assess the overall scientific quality of the program and the extent to which the research it supports has led to technical or intellectual "breakthroughs" of value to the scientific community and technology development efforts.

The committee recommends that the independent review panel be charged with the responsibility of assessing the quality of EMSP science and its impacts. The committee recommends that the Program Director assume the responsibility for developing a "portfolio" of information that would support both shorter-term and long-term assessment of EMSP by the independent review panel.

APPLYING RESULTS OF BASIC RESEARCH
TO THE CLEANUP MISSION

The movement of new knowledge and insights from investigators to full-scale application is a slow and diffuse process. **As a way of facilitating this information flow and stimulating new research ideas, the EMSP Program Director should convene annual workshops, seminars, and symposia that bring together EMSP investigators, program managers from EM and ER (including those in the EM focus areas), site contractors and other problem holders, and, when appropriate, other stakeholders, regulators, and P.I.s and managers from other research programs. The Program Director should assume responsibility for determining how to best structure such activities so that they serve the interests of investigators and EM's needs for information transfer.** It will be important in any effort that is undertaken to improve communication and information flow to involve the problem holders at the sites. These individuals will not only have the greatest knowledge about the sites but will also be able to assist in integrating the results of EMSP into the long-term EM effort.

The responsibility for disseminating results from EMSP is not EMSP's alone. Other offices in EM, especially other parts of the Office of Science and Technology, must take an active role in ensuring that the Department and the nation reap the full benefits from EMSP-supported research. Without an active effort to move research into technology development and application, the EMSP may become a high-quality research program but have little limited impact on the EM's cleanup mission.

1
INTRODUCTION AND BACKGROUND

The Department of Energy's (DOE's) Environmental Management Science Program (EMSP) was created by the 104th Congress to stimulate basic research and technology development for environmental cleanup of the nation's nuclear weapons complex. The program was created in the conference report that accompanied the Energy and Water Development Appropriations Bill:[1]

> The conferees agree with the concern expressed by the Senate that the Department [of Energy] is not providing sufficient attention and resources to longer term basic science research which needs to be done to ultimately reduce cleanup costs. The current technology development program continues to favor near-term applied research efforts while failing to utilize the existing basic research infrastructure within the Department and the Office of Energy Research. As a result of this, the conferees direct that at least $50,000,000 of the technology development funding provided to the environmental management program in fiscal year 1996 be managed by the Office of Energy Research and used to develop a program that takes advantage of laboratory and university expertise. This funding is to be used to stimulate the required basic research, development and demonstration efforts to seek new and innovative cleanup methods to replace current conventional approaches which are often costly and ineffective.

The DOE Office of Environmental Management (EM)—which is responsible for cleanup of the nation's nuclear weapons complex—and

---

[1]H.R. 1905, which was enrolled as Public Law 104-46, 1995.

the DOE Office of Energy Research (ER)—which manages the Department's basic research programs—formed a partnership to develop a mission-directed basic research program focused on EM's long-term cleanup problems. According to the fiscal year 1996 (FY96) announcement for this program,[2] the objectives of the EMSP are to

- Provide scientific knowledge that will revolutionize technologies and clean-up approaches to significantly reduce future costs, schedules, and risks;
- "Bridge the Gap" between broad fundamental research that has wide-ranging applicability such as that performed in DOE's Office of Energy Research and needs-driven applied technology development that is conducted in EM's Office of Science and Technology; and
- Focus the Nation's science infrastructure on critical DOE environmental management problems.

The FY96 program announcement invited investigators from universities, industry, and national laboratories to submit research ideas to the Department in the form of brief preproposals. The program received 2,149 preproposals in response to these announcements. These preproposals were reviewed by federal program managers, and the proposers of 793 projects were encouraged to submit full proposals. A total of 810 full proposals were received, covering a wide range of disciplines and research topics. The Department convened review panels to evaluate the scientific/technical merit of the proposals and their long-term relevance to EM's cleanup mission and used the advice of these panelists to make 140 three-year awards totaling about $112 million.

The Committee on Building an Environmental Management Science Program was established under the auspices of the National Research Council at the request of Thomas P. Grumbly, Under Secretary of Energy, to advise the Department on the structure and management of the EMSP. The committee met seven times during the period May to

---

[2]The solicitation to university and industry researchers was published in the *Federal Register* on February 9, 1996. A similar solicitation was provided to national laboratory researchers at about the same time.

November 1996 and produced three reports: an Initial Assessment Report[3] (Appendix F) that addresses the near-term needs of the program related to the FY96 proposal competition; a Letter Report[4] (Appendix G) that addresses the development of an FY97 program announcement; and the present report, which addresses longer-term challenges and opportunities for the program.

INITIAL ASSESSMENT REPORT

The committee's Initial Assessment Report (Appendix F) was released on July 9, 1996—just in time for the Department's use in decision making on awards in the FY96 proposal competition. The report provided a brief review of the DOE cleanup mission and provided comments on the value of basic research to that mission. The report also provided comments on several program "challenges," in particular, challenges related to attracting outstanding investigators to the program, obtaining innovative research, and applying the results of this research to the cleanup mission.

The Initial Assessment Report's findings and recommendations were preliminary in nature, but many bear repeating in this final report and are summarized below:

• Given the size, scope, and long-term nature of DOE's cleanup mission,[5] the committee views the establishment of the EMSP as a prudent and urgent investment for the nation. The nation's first-year financial investment in the EMSP—$50 million—is modest compared to the Department's $6.1 billion annual investment in cleanup.[6]

---

[3]National Research Council, 1996, Building an Effective Environmental Management Science Program: Initial Assessment (Washington, D.C.: National Academy Press). This report is available on the World Wide Web at the following address: http://www.nap.edu/readingroom/books/envmanage/index.html.

[4]Letter Report to the Associate Deputy Assistant Secretary for Science and Risk Policy, October 8, 1996.

[5]The Department estimated in early 1996 that this effort will cost between about $190 billion and $265 billion and require several decades to complete. The "10-year vision" of the Assistant Secretary for Environmental Management has the objective of accelerating much of this effort. See Chapters 2 and 3.

[6]Funding for the EMSP represents about 0.6 percent of EM's annual budget.

• Many of the nation's better scientists and graduate students have the backgrounds and skills necessary to do work at the forefront in this area but are not currently involved in research of relevance to the EMSP. The Department will need to demonstrate a long-term commitment to this research program before some of these scientists will redirect their research and graduate student training activities to the program's concerns. To this end, the committee recommended that the Department take the following three actions: (1) maintain at least level funding for the program in FY97; (2) provide full funding for all approved projects in the FY96 competition out of FY96 funds;[7] and (3) relax the initial allocation[8] of $20 million for proposals from academia and industry and $20 million for proposals from national laboratories to the extent allowed by the law and, instead, support the most scientifically meritorious and relevant work, regardless of the institution of origin.

• To obtain highly meritorious research proposals, the Department must help investigators become more knowledgeable about its cleanup problems, both generic problems and site-specific problems. To this end, the committee recommended that DOE prepare concise technical summaries of the critical barriers to the solution of cleanup problems with the advice of the research and research-user communities.

• For the EMSP to contribute to the long-term cleanup mission, effective mechanisms must be found to transfer the results of research sponsored by the EMSP to potential "users" in government, industry, and academia who can utilize this knowledge to develop new or improved cleanup methods.

The reception of the committee's Initial Assessment Report by the Department was positive, and Department staff moved expeditiously to implement many of the committee's recommendations. Most notably, the Department relaxed its initial allocation of funding for university/industry and national laboratory proposals and instead made funding decisions based on merit and relevance. The Department awarded about $43 million to university/industry projects and $69

---

[7] The Department made three-year awards for projects funded in the FY96 competition. The committee recommended that funding for all three years be provided out of FY96 funds so that the Department would not have to "mortgage" funds for this program in subsequent years.

[8] As stated in the FY96 program announcement.

million to national laboratory projects.[9] The Department provided full funding for university proposals but did not fully fund proposals from national laboratories, arguing that it was unable to do so.

The Department's FY97 budget request had already been submitted to Congress when the committee issued its Initial Assessment Report. However, the Congress increased the funding for this program from the Department's request of $38 million to $50 million, noting that[10]

> The conferees are pleased with the progress to date in implementing the environmental basic research program. In a recent review by the National Research Council, the Council endorsed this program and acknowledged, '* * * establishment of this mission-directed, basic research program as both an urgent and a prudent investment for the nation.' The National Research Council report further notes that the, '* * * long-term success of this program is highly dependent on the continuing partnership between EM, which understands the cleanup problems and research needs, and ER, which, through its mission to manage the department's basic research programs, understands how to select and manage research. The committee endorses the efforts made by EM and ER staff to work together and encourages them to continue their efforts to build an effective Environmental Management Science Program.'

LETTER REPORT

The committee also recommended in. its Initial Assessment Report that the Department postpone the release of the FY97 program announcement until it had more time to identify and incorporate "lessons

---

[9]A total of $47 million was provided out of FY96 funds, $43 million to university and industry researchers and $4 million to national laboratory researchers. The balance of funding to national laboratory researchers—$65 million—will be obtained from future-year congressional allocations to the program.

[10]The text is taken from the Conference Report on H.R. 3816, Energy and Water Development Appropriations Act, 1997.

learned" from the FY96 program competition. This recommendation prompted the Department to request additional advice from the committee on the FY97 program announcement,[11] which in turn led to the production of a Letter Report. This Letter Report (Appendix G) provided an initial assessment of the FY96 proposal competition and offered advice on several aspects of the FY97 program announcement. The committee concluded that it did not have enough time, nor was enough information provided, to assess the overall success of the FY96 competition. However, the committee noted that, where it did have firsthand information, it was able to confirm the overall quality of the proposals, the review process, and the review panelists.

The Letter Report also offered suggestions on several elements of the FY97 program announcement, most notably the following:

- *Criteria for proposal review and selection.* The committee recommended a continued focus on basic research in the program announcement, with scientific merit and long-term relevance to EM's cleanup mission as the primary proposal screening criteria.
- *Research areas.* The committee suggested that the EMSP would be more likely to attract innovative proposals from creative investigators if the program announcement contained information on EM's problems that require basic research. The committee also encouraged the Department to broaden the solicitation to include problems related to risk, quantitative methodologies, and health assessment.
- *Review process.* The committee reaffirmed its endorsement of the two-stage review process—consisting of a scientific and technical merit review followed by a relevance review—and recommended that this process continue to be managed as a partnership between ER and EM. In addition, the committee recommended that the Department maintain some continuity in the merit and relevance review panels to take advantage of the experience gained in the FY96 competition. The committee also recommended that ER convene the merit review panels under the Federal Advisory Committee Act (FACA) to allow the

---

[11]Written request from the Associate Deputy Assistant Secretary for Science and Technology dated August 9, 1996.

panelists to provide formal consensus on scoring and ranking of proposals to DOE.

• *Financial plan.* The committee expressed its concerns about the "mortgage" from the FY96 proposal competition[12] and reaffirmed the recommendation in the Initial Assessment Report that successful proposals be fully funded "up front."

• *Outreach.* The committee encouraged the Department to explore additional mechanisms to make the research community more broadly aware of the FY97 proposal competition, for example, through the use of paid advertisements in professional journals.

At the time the present report was prepared, the Department had not released its FY97 program announcement; consequently, the committee was not able to determine the extent to which its advice was followed. The committee has received informal feedback from Department staff that suggests that many of the recommendations will be implemented, with the exception of full funding for national laboratory proposals and the use of FACA panels for merit review, which will not be implemented unless certain institutional obstacles are overcome. The committee provides some additional comments on these issues later in this report.

FOCUS OF THIS REPORT

The purpose of the present report is to provide a more detailed assessment of the EMSP than was possible in the committee's Initial Assessment Report, which was prepared on a very tight schedule and with the benefit of only two committee meetings. The primary focus of the present report is on the long-term challenges and opportunities for the program as noted in the Statement of Task, which is given in Appendix A.

---

[12]This mortgage consists of future-year funding commitments to national laboratory researchers because the Department did not provide full funding for FY96 proposals from FY96 program funds. As shown in Attachment E of the Letter Report (and Table 4.2 of this report), this mortgage includes commitments of $23 million in FY97, $23 million in FY98, and $19 million in FY99. These commitments will reduce substantially the funds available to support new or competitive renewal proposals in future years unless the budget for the program is increased.

The chapters of this report have been structured to address the Statement of Task questions in Appendix A. Chapter 2 addresses the value of basic research to the cleanup program, recapitulating and extending the comments made by the committee in its Initial Assessment Report and Letter Report. Chapter 3 addresses the five questions under "science needs" in the Statement of Task in the context of the development of a *science plan* for the program. Chapters 4 and 5 address the four questions under "management needs" for the program. Chapter 4 deals with proposal selection and funding, whereas Chapter 5 addresses the structure and operation of the program. In addressing its task statement the committee focused on the "big picture" issues that are likely to be of value to the Department, recognizing that the details are best left to program staff.

LIMITATIONS OF THIS REPORT

The committee was not able to address comprehensively all of the task statements for the following two reasons:

1. The committee worked to a series of deadlines set by the FY96 proposal competition and FY97 program announcement processes. The committee was able to affect the initial EMSP program only by producing its reports in a much more rapid fashion than is usual for NRC studies. The committee has been successful in providing guidance to the initial program but has not gone into depth in some areas, most notably the selection of specific research areas for emphasis, because this task would have exceeded the time allotted to the committee for the completion of its work and possibly its expertise.

2. The EM program itself lacks clear objectives, including what will be the land uses at the DOE sites and what the standards are by which "cleanup" will be judged to be completed. These issues have been addressed by other reports,[13] which indicate that DOE has expended large

---

[13]For example, General Accounting Office, 1994, Nuclear Cleanup: Completion of Standards and Effectiveness of Land Use Planning Are Uncertain, GAO/RCED 94-144 (Washington, D.C.: GAO); DOE, Secretary of Energy Advisory Board, 1995, Alternative Futures for the Department of Energy National Laboratories, SEAB—95006873 (Washington, D.C.: DOE); National Research Council, 1996, Improving the

amounts of funds but accomplished little. The new 10-year vision plan of the Assistant Secretary for Environmental Management, which is discussed in Chapter 3, is based on setting objectives. However, this plan acknowledges that the most difficult problems, dealing with transuranic (TRU) waste and high-level waste (HLW),[14] will not be resolved in the 10-year period. This lack of objectives in the EM program itself, in the view of some committee members, is a serious flaw in trying to develop a needs-based basic research program.

SOURCES OF INFORMATION FOR THIS REPORT

The committee obtained a wide range of oral briefings and written documentation during this study. A list of briefings received at the committee's open meetings is given in Appendix B. The committee received several detailed briefings from EM and ER program staff on the structure and management of the EMSP, proposal review and award procedures, and results of the FY96 proposal competition. The committee also received extensive written documentation from the Department, including a multi-volume record and a data table that provided a list of all projects funded in the FY96 proposal competition that included principal investigator (P.I.) and co-P.I. names and affiliations, biographical sketches of the P.I.s, abstracts of funded projects, and current DOE funding. Additionally, the committee received in confidence about half of the names of the FY96 relevance review panelists from EM.

The committee received oral briefings from staff at DOE, the National Science Foundation (NSF), the U.S. Environmental Protection Agency (EPA), and the National Institutes of Health (NIH) on other federal programs of relevance to the EMSP. The committee also received oral briefings from federal and private-sector managers on effective R&D program management and assessment strategies.

In its efforts to understand the cleanup challenges at the weapons complex, the committee solicited and received an extensive set of

---

Environment: An Evaluation of DOE's Environmental Management Program (Washington, D.C.: National Academy Press).

[14]Transuranic waste contains nuclides such as plutonium that have atomic numbers greater than 92 (uranium). High-level waste is highly radioactive material that contains fission products and transuranic elements. Both types of waste are generated during reprocessing of irradiated fuel for plutonium production.

briefings over two meetings from DOE, contractor, and national laboratory staff. The first set of briefings reviewed cleanup problems and R&D needs at five of the largest sites—Hanford, Savannah River, Rocky Flats, Idaho Falls, and Oak Ridge. The second set of briefings reviewed cleanup problems arranged by focus area—landfills and plumes, tanks, mixed wastes, and decontamination and decommissioning. The committee found these briefings to be helpful in clarifying its thinking about the need for basic research in the weapons complex.

During the course of this study, the committee made numerous and significant requests for information from Department staff. The committee found the staff to be responsive to requests for information, with one exception as noted below, and the committee generally was satisfied with the quality and completeness of the information it received and the willingness of Department staff to respond in a timely fashion. The committee also was pleased by the candor of Department staff in discussing program problems and their willingness to engage in wide-ranging and vigorous discussions of the program at the committee's open sessions.

The one exception involved the committee's request for the names of the merit review panelists from ER staff. The committee requested these names as part of its efforts to assess the quality of the peer review in the FY96 proposal competition (see Chapter 4) but was told that it was ER practice to keep the names confidential. After discussions with ER staff over the course of three committee meetings, ER staff agreed to contact the panelists to see if they would agree to have their names released to the committee. The panelists had not been contacted by the time of the committee's penultimate meeting.

2

VALUE OF THE EMSP
TO THE CLEANUP MISSION

The Department of Energy's (DOE's) Office of Environmental Management (EM) is responsible for cleanup of the nation's nuclear weapons complex, a vast network of industrial sites established during the Second World War and the Cold War to develop, test, and produce nuclear weapons.[1] The EM cleanup mission is massive in scope: it includes 3,700 contaminated sites in 34 states and territories; more than 100 million gallons of radioactive and mixed wastes stored in 322 tanks; 3 million cubic meters of radioactive or hazardous buried wastes; 250 million cubic meters of contaminated soils from landfills and plumes; more than 600 billion gallons of contaminated ground water; and about 1,200 facilities that require decontamination and decommissioning.[2] The Department estimates[3] that cleanup of the weapons complex will cost between about $190 billion and $265 billion and take several decades to complete; these estimates do not include costs for dealing with "intractable" problems such as the large volumes of contaminated soil and ground water around the complex.[4]

Many of EM's cleanup problems cannot be solved or even managed efficiently and safely with current technologies, in part owing to their

---

[1]DOE, 1995, Closing the Circle on the Splitting of the Atom: The Environmental Legacy of Nuclear Weapons Production in the United States and What the Department of Energy Is Doing About It (Washington, D.C.: DOE).

[2]From written material received from EM at the first committee meeting, May 11-12, 1996.

[3]DOE, 1996, Estimating the Cold War Mortgage: The 1996 Baseline Environmental Management Report, DOE/EM-0290, 3 vols. (Washington, D.C.: DOE).

[4]In parallel with this committee's efforts, the Assistant Secretary for Environmental Management recently described a "10-year vision" plan through which he intends to focus and accelerate cleanup of the weapons complex. The details of the plan have not yet been made public, but preliminary descriptions recognize that many of the more difficult problems will take longer than 10 years to resolve with current technical understanding. The Department plans to release a draft of the plan during the first quarter of 1997.

tremendous size and scope. However, cleanup would benefit greatly from the involvement of basic researchers, as noted in a recent NRC report:[5]

> In some circumstances, technologies and processes for safe and efficient remediation or waste minimization do not exist. In other cases, the development of new technology and processes might substantially reduce the costs of, or risks associated with, remediation and waste management. . . . In some cases, fundamental science questions will have to be addressed before a technology or process can be engineered. . . . There is a need to involve more basic science researchers in the challenges of the Department's remediation effort.

The importance of basic research to the EM cleanup mission was highlighted in the report of the Task Force of Alternative Futures for the Department of Energy National Laboratories,[6] also known as the Galvin report. The report (p. 6) noted that

> the Department faces a monumental task in dealing with the radioactive and hazardous wastes at its former nuclear weapons production sites and national laboratories. This task cannot be addressed in an affordable fashion using today's technologies.

The report (pp. 40-41) further asserts that

> there is a particular need for long-term, basic research in disciplines related to environmental cleanup. . . . Adopting a science-based approach that includes supporting development of technologies and expertise . . . could lead both to reduced cleanup costs and smaller environmental

---

[5]National Research Council, 1996, Improving the Environment: An Evaluation of DOE's Environmental Management Program (Washington, D.C.: National Academy Press).

[6]DOE, Secretary of Energy Advisory Board, 1995, Alternative Futures for the Department of Energy National Laboratories, SEAB—95006873 (Washington, D.C.: DOE).

impacts at existing sites and to the development of a scientific foundation for advances in environmental technologies.

The National Research Council called for a closer linkage between basic research and technology development in EM:[7]

> EM has recently begun an effort to coordinate its technology development efforts with the Office of Energy Research, which houses much of the Department's basic research and is the principal office for interaction with nondefense Department National Laboratories. . . . This type of linkage, including the defense-related laboratories, where much of the expertise in nuclear materials resides, is precisely what is called for. . . . The Department should extend this attempt to create partnerships to include the basic-research efforts in universities and industrial concerns that are developing technology or undertaking their own research.

The committee agrees with these assessments and believes that a basic research program focused on EM's most difficult cleanup problems may have a significant long-term impact on the EM mission. Basic research may provide new knowledge to allow the Department to attack cleanup problems that are currently intractable or exorbitantly expensive using current technologies; it may lead to the development of better technologies to allow cleanup to be accomplished at lower costs or with fewer hazards to workers and the public; it can improve understanding of risks, and how to discuss them with local stakeholders; and it may lead to the development of new or improved technologies that will allow cleanup to a higher state than is presently possible, thereby making sites available for less restrictive uses. Simply put, new technologies are required to deal with EM's most difficult problems, and new technologies demand new science.

---

[7]National Research Council, 1996 (see footnote 5), p. 117.

A basic research program designed to address fundamental principles may lead to discoveries that change present ways of thinking and lead to more powerful scientific paradigms. Creative investigators are drawn to basic research by the challenge of solving interesting problems in science. There certainly is no shortage of interesting problems related to EM's mission. Indeed, the committee believes that a basic research program focused on EM's problems could transcend the EM program and be useful in the much larger scientific and environmental arenas. Such broad applicability is a typical outcome of basic research.

Basic research already has helped in the cleanup effort. For example,

1. Basic research on the kinds of chemical and biological reactions that transform pollutants has led to treatment approaches for contaminants. For example, many organic contaminants that exist at DOE sites (chlorinated solvents and hydrocarbons) can be biodegraded. This has led to great interest in the use of bioremediation for control of contaminated ground waters and soils.[8] Bioremediation can be used in many ways, including biodegradation of concentrated petroleum hydrocarbon contaminants near their source, biodegradation of dilute contaminants in large plumes, removal of residual contaminants following physical or chemical methods, and for capture of metals and radionuclides through microbially mediated transformation processes.[9] Numerous case studies of bioremediation treatment systems are published in the *Proceedings of the Third International In Situ and On-Site Bioreclamation Symposium.*[10]

Another strategy under development for controlling the risks associated with contaminated subsurface environments is to use hydraulic

---

[8]National Research Council, 1993, In Situ Bioremediation, When Does It Work? (Washington, D.C.: National Academy Press).

[9]National Research Council, 1994, Alternatives for Ground Water Cleanup (Washington, D.C.: National Academy Press); R. D. Norris, R. E. Hinchee, R. Brown, P. L. McCarty, L. Semprini, J. T. Wilson, D. H. Kampbell, M. Reinhard, E. J. Bouwer, R. C. Borden, T. M. Vogel, J. M. Thomas, and C. H. Ward, 1994, Handbook of Bioremediation (Boca Raton, Fla.: CRC Press).

[10]Proceedings of the Third International In Situ and On-Site Bioreclamation Symposium, Volumes 1-10, 1995 (Columbus, Ohio: Battelle Press).

barriers to direct contaminated ground-water flow through a reactive medium (NRC, 1994; see footnote 9). The reactive medium can use a combination of physical, chemical, and biological processes. A zero-valent iron permeable barrier has shown promise for the removal of highly chlorinated solvents such as tetrachloroethene and trichloroethene.[11] The concept of a permeable treatment barrier is being evaluated for treatment of metals and radionuclides.[12] Researchers have discovered anaerobic bacteria that rapidly reduce uranium(VI), which is soluble, to uranium(IV), which precipitates.[13] Thus, it may be possible to immobilize uranium and other radionuclide contaminants, such as plutonium and technetium, by stimulating microbial reduction of the metal in the reaction zone of a permeable barrier.

The improvement in understanding the role of physical, chemical, and biological processes in the fate of contaminants has led to a big change in the way ground-water cleanups are now being approached and carried out (with tremendous cost savings). There is much interest now in determining if the natural processes are sufficient to keep the risk low and serve as a "protective barrier" to prevent excessive migration of contaminants. It is difficult to consider the approach of natural attenuation or intrinsic remediation without a firm understanding of the basic physical, biological, and chemical reactions.

2. Some contaminants that are known to strongly sorb to soil have been observed to migrate great distances with ground-water flow. In this situation the contaminant was thought not to be a problem, but the excessive migration means there is an elevated risk. It has been found that the mobility is due to transport of contaminants bound to colloids, not the chemical moving by itself. Knowledge of the behavior of colloids in ground water has led to explanations for why some contaminants can migrate great distances in ground water. This is an example of how basic research on colloid movement and interaction with contaminants has improved our characterization of the risk.

---

[11]E. K. Wilson, 1995, Zero-Valent Metals Provide Possible Solution to Groundwater Problems, Chemical and Engineering News 73(27):19-22.

[12]Department of Energy, 1996, Subsurface Contaminants Focus Area Technology Summary, Office of Science and Technology, DOE/EM-0296, pp. 142-144.

[13]D. Lovley and E. J. P. Phillips, 1992, Reduction of uranium by *Desulfovibrio desulfuricans*, Applied and Environmental Microbiology, 58(3):850-856; D. Lovley, E. J. P. Phillips, Y. A. Gorby, and E. R. Landa, 1991, Microbial Reduction of Uranium, Nature 350:413-416.

In its Initial Assessment Report the committee suggested that a basic research program could produce knowledge that, if properly applied in technology development efforts, could address the following EM challenges:

- *Characterization, remediation, and management of radioactive and chemical wastes.* Basic research may help stimulate the development of new technologies and reduce the uncertainties involved in the application of current technologies.
- *Secondary wastes.* Basic research may lead to the development of new methods to reduce the volume and toxicity of the secondary wastes generated during cleanup.
- *Risk.* Basic research may provide a better understanding of risk, which would help EM prioritize its cleanup activities and reduce hazards to workers, the public, and the environment.

The Environmental Management Science Program (EMSP), as currently designed, is a *needs-driven* or *mission-directed basic research* program: *needs-driven* in the sense that research is focused in areas where new knowledge may stimulate the development of new technologies for cleanup, and *basic* in the sense that the program supports research on fundamental processes and phenomena with no specific outcome or time horizon for application. The EMSP is designed to feed into a much larger technology development program within EM.[14]

The EMSP is different in several respects from other federal basic research programs, including other DOE programs, that support fundamental research related to hazardous materials in the environment and environmental management (e.g., Table 3.1). Although several federal programs support basic research in fields broadly relevant to environmental science, none focus explicitly on EM's problems, and none have an explicit link to the problem holders at the sites that the committee recommends be established (see Chapter 5).

In addition to its value for generating new knowledge, the EMSP supports EM's mission in two other important ways. First, the program contributes to training future generations of scientists and engineers—an

---

[14]Technology development efforts within EM are managed by the Office of Science and Technology (EM-50), which has an annual budget (excluding the EMSP) of about $316 million per year in FY 1997.

important consideration for any agency or program with a mission that will last several decades. This training will secure future access to knowledge long after the current generation of investigators has passed from the scene. This training also may promote the development of what the committee has referred to in its previous reports as a "committed cadre" of investigators for the program—scientists knowledgeable about EM's problems and needs who produce knowledge of long-term value to the cleanup mission.

Second, the EMSP will promote the development of partnerships among universities, national laboratories, other federal agencies, and the private sector. These partnerships bring together highly creative and innovative researchers, provide access to unique national research facilities, and provide a multidisciplinary focus on EM's most difficult problems.[15]

As the committee stated in its Initial Assessment Report, funding for the EMSP should be viewed as an investment that may, in the long term, lead to more effective cleanup. This investment in basic science is not unlike the R&D investments made by successful for-profit, private-sector firms, which recognize that R&D is essential to long-term success.

The committee reiterates that EMSP is not a "cure-all"—it alone will not solve all of EM's cleanup problems. As with any basic research program, there are no guarantees of quantifiable returns, and, indeed, it may be difficult to track precisely the returns on dollars invested. However, the sheer magnitude of the cleanup mission and its estimated cost, coupled with the technological challenges, make the investment in EMSP both prudent and timely, even urgent. The committee believes that basic research will lead to new knowledge which, given the scope and time frame of the problem, will be of value to cleanup of the weapons complex. There is no way to forecast the impact of carefully chosen high-quality projects, but the EM problems are so large and the predicted costs of cleanup so great, that the modest investment in the EMSP is viewed as worthwhile. In the committee's view, the potential benefits of the program clearly justify its continued support.

---

[15]Collaborations are discussed further in Chapter 3.

3

EMSP SCIENCE PLAN

The statement of task for this report (Appendix A) directed the committee to address five questions related to "science needs" for the Environmental Management Science Program (EMSP). Broadly speaking, the committee was asked to provide advice on an EMSP research agenda, both in terms of process and content. The committee also was asked to provide advice on how the EMSP could best leverage its research investments and broaden the community of investigators available to address problems of concern to the program. Some of these questions were addressed in earlier committee reports, as noted in Chapter 1. In this chapter the committee summarizes the conclusions from its previous reports and provides additional advice on the following issues:

- rationale for developing a science plan for the EMSP,
- content of and process for developing the science plan,
- strategies for coordinating the investment in basic research, and
- strategies for broadening the investigator community involved in work of relevance to the EMSP.

RATIONALE FOR DEVELOPING A
SCIENCE PLAN FOR THE EMSP

The overall goals of the Environmental Management Program (EM) have been under formulation for at least seven years, since the program began under Secretary Watkins. Amidst the many pressures of local stakeholders, regional Environmental Protection Agency (EPA) officials, and state environmental officials, the Department of Energy (DOE) has been trying to establish a program that will enable it to deal with the legacy of the Cold War weapons production facilities. The DOE has called this a cleanup program but has not attempted to define what cleanup is. The most recent attempt to describe what cleanup might entail was made by Assistant Secretary Alm, who has proposed a 10-year vision for resolving many issues of concern to local stakeholder groups at the sites. He explicitly acknowledges that some of the problems, notably transuranic waste, high-level waste, and ground-water contamination, will not be resolved in a 10-year time frame. Because of local conditions, especially the views of local stakeholder groups, the DOE has not tried to establish a single national level-of-cleanup standard.

With this as background, it should not come as a surprise that the committee had a great deal of trouble addressing the following question in its Statement of Task: "What areas of basic research are likely to provide the best payoffs for EM cleanup efforts over the next few decades?" Indeed, after extensive discussions and many presentations from DOE representatives, contractors, and national laboratory staff, the committee concluded that it could not provide an explicit answer to this question without many more meetings and perhaps a different committee membership. A majority of the committee believes that, because the EMSP is so new and represents a different way of approaching the cleanup problems, it may not even be wise to make detailed recommendations with regard to the inclusion or exclusion of specific research areas. The research content of the EMSP will likely evolve over time as results are accumulated and evaluations of outcomes for the broader EM effort are conducted. The committee did conclude, however, that it could recommend a process that DOE could follow to identify its research needs, and it focuses on that process in the following section.

The Congress's rationale for creating the EMSP developed from a sense that DOE was not devoting sufficient time or resources to fundamental scientific studies that would be of benefit to cleanup in the long term. As the Congress noted in the conference report on the Energy and Water Development Appropriation Bill that created the EMSP, ". . . the Department [of Energy] is not providing sufficient attention and resources to longer term basic science research which needs to be done to ultimately reduce cleanup costs."[1] Indeed, the committee's review of some of the Department's documentation of cleanup needs and strategies reinforces the impression that the Department itself has not acknowledged the need for or the potential value of basic research in its cleanup mission.

---

[1]See Chapter 1, page 1 for a more complete quotation.

The Government Performance and Results Act (GPRA) requires that by 30 September 1997 each agency submit to the Office of Management and Budget (OMB) and the Congress a strategic plan that contains, among other items, "a comprehensive mission statement covering the major functions and operations of the agency" and "a description of how the goals and objectives are to be achieved. . . ."[2] Many agencies are moving to develop such plans in advance of the required date. In February 1996, for example, the National Aeronautics and Space Administration (NASA) released its strategic plan.[3] In May 1996, EPA published a strategic plan for the Office of Research and Development and a science planning document.[4]

DOE has begun to produce related documents. In July 1996, the Department published the first part of its strategic plan for national laboratories.[5] In August 1996, the Department released its strategic plan for energy research.[6] As mentioned in Chapter 2, the new Assistant Secretary for Environmental Management has begun an ambitious program to develop a strategic plan related to his 10-year vision.[7] The Department also has produced a report that details its plans for land and infrastructure use at 20 DOE sites.[8] Another congressionally requested

---

[2]The quoted text is from the GPRA, 5 U.S.C. Section 306(a)(1) and 306(a)(3). The reference to the OMB document for strategic plans is Office of Management and Budget, 1996, Preparation and Submission of Strategic Plans, Circular No. A-11, Part 2 (Washington, D.C.: OMB).

[3]NASA. 1996. NASA Strategic Plan (Washington, D.C.: NASA).

[4]EPA, 1996, Strategic Plan for the Office of Research and Development, EPA/600/R-96/059 (Washington, D.C.: EPA); EPA, 1996, Report to Congress: The Science to Achieve Results (STAR) Program, EPA/600/R-96/064 (Washington, D.C.: EPA).

[5]DOE, Laboratory Operations Board, 1996, Strategic Laboratory Mission Plan—Phase I (Washington, D.C.: DOE).

[6]DOE, 1996, Energy Research Strategic Plan, DOE/ER-0656 (Washington, D.C.: DOE).

[7]DOE, 1996, Memo, 10 June 1996, from Assistant Secretary for Environmental Management regarding Integrated Strategic Planning, Budgeting and Management System/10 Year Planning.

[8]DOE, 1996, Charting the Course: The Future Use Report, DOE/EM-0283 (Washington, D.C.: DOE).

report, *The 1996 Baseline Environmental Management Report,*[9] usually referred to as the BEMR, estimates life-cycle costs and schedules for completing EM's mission. Finally, there are documents published by the EM Office of Science and Technology, for example, its annual report to Congress[10] and reports from the focus areas.[11]

To the committee's knowledge, the Department has not explained the role of the EMSP in the cleanup mission in any of these planning documents. For example, neither the BEMR Executive Summary nor the ER Strategic Plan mentions the EMSP. The Strategic Laboratory Mission Plan presents a volume of mission activity profiles: 52 for national security, 53 for energy resources, 54 for science and technology, and 7 for environmental quality. Of these seven, one is on Yucca Mountain, two are on storing or removing spent fuel from commercial reactors, one on developing an integrated waste management system by the Office of Civilian Radioactive Waste Management, one is on field support for West Valley and the Waste Isolation Pilot Plant (WIPP), one is "directed toward satisfying compliance agreements and regulatory requirements," and one is on technology development, essentially the Office of Science and Technology (EM-50). None address the basic research that is the theme of the EMSP.

Indeed, given the near-term budgetary and scheduling pressures on the program—EM is being urged to "get on" with cleanup from an impatient Congress and public while at the same time its budget and staff are under significant downward pressures—the Department has had little opportunity or incentive to advocate long-term investments in scientific research, a position reinforced by the Assistant Secretary's 10-year vision: "Within a decade, the EM program will complete cleanup at most nuclear sites." The implicit "message" of this vision is that most of the

---

[9]DOE, Office of Environmental Management, 1996, The 1996 Baseline Environmental Management Report, DOE/EM-0290, 3 vols. and Executive Summary (Washington, D.C.: DOE).

[10]DOE, Office of Science and Technology, 1996, Annual Report to Congress, FY1995 (Washington, D.C.: DOE).

[11]DOE, 1996, Characterization, Monitoring and Sensor Technology Crosscutting Program: Technology Summary, DOE/EM-0298 (Washington, D.C.: DOE). The Department has developed other focus area reports on the following topics: subsurface contaminants; decontamination and decommissioning; plutonium; robotics; mixed waste characterization, treatment and disposal; radioactive tank waste remediation; and efficient separations and processing.

cleanup of the weapons complex can be completed in 10 years using currently known technology and understanding. However, the Assistant Secretary does recognize that cleanup will not be completed in 10 years: "At a small number of sites treatment will continue for the few remaining waste streams. . . . Remaining waste streams include high-level and TRU [transuranic] wastes."[12] Thus, many of the most difficult problems will remain even if the 10-year plan is successful. It is just these types of problems that will require the results of the EMSP.

As noted in a previous chapter and in its Initial Assessment Report, the committee finds good reasons for long-term investments by EM in basic scientific research that is not linked to the 10-year vision: these investments can provide new knowledge that will allow the Department to attack cleanup problems that are currently intractable or exorbitantly expensive using current technologies; they can lead to the development of better technologies to allow cleanup to be accomplished at lower costs or with fewer hazards to workers and the public; and they can lead to the development of new or improved technologies that will allow cleanup to a higher state than is presently possible, thereby making sites available for less restrictive uses.

If the EMSP is to have a significant impact on the cleanup mission, the Department must incorporate this program into its strategic plans. Indeed, as the deadline for GPRA's reporting requirements draws near, it is essential to the survival of the EMSP that a plan for applying basic research in the cleanup program—a *science plan*—be explicitly and officially articulated by the Department. **As a first step to this end, the committee recommends that the Department develop a science plan for the EMSP. This science plan should provide a comprehensive list of significant cleanup problems in the complex that can be addressed through basic research and a strategy for addressing them.**

A majority of the committee members believe that basic research focused on EM's more difficult cleanup problems (which are understood in a broad sense) may provide the information necessary for the cleanup program to succeed. However, one member of the committee believes that the lack of clear objectives for the cleanup program requires the

---

[12]DOE, 1996, Memo, 10 June 1996, from Assistant Secretary for Environmental Management regarding Integrated Strategic Planning, Budgeting and Management System/10 Year Planning, pp. 1-2 of attached Draft Guidance for the 10-Year Plan.

EMSP and the relevance reviews to "fly blind." This member argues that the science plan should further the objectives of the EM program (whatever they are determined to be). Without clear objectives for the EM program, this member sees a logical flaw in recommending the development of such a plan. The majority of the committee disagrees and believes that enough is known about the contamination problems at DOE sites that the development of a science plan will improve the EMSP. However, all members agree, as other National Research Council panels have, that DOE should place greater emphasis on defining a set of specific near- and long-term objectives for the cleanup program.

CONTENT OF AND PROCESS FOR DEVELOPING THE SCIENCE PLAN

The committee views the science plan as the primary guiding document for the Department's basic research investment in the cleanup mission. To serve this purpose, the content of the science plan needs to be comprehensive and reflective of the significant cleanup problems in the complex. The committee's Letter Report encouraged the Department to broaden its research solicitations and to include problems related to risk, health assessment, and quantitative methodologies (i.e., statistical methods, numerical [simulation] methods and the combination of the two sets of techniques), mainly because the committee believes that research in these areas could have a direct impact on the cleanup mission (e.g., Sidebar 3.1). In addition, the committee believes that ER should ensure that the pertinent merit review panelists are knowledgeable in the risk research field.

The committee addressed the identification of cleanup problems in its first two reports.[13] In its Initial Assessment Report, the committee recommended "that DOE prepare concise written technical summaries of its basic research needs for the research community. Such summaries should contain information on the critical barriers to the solution of EM's problems, arranged both by site and by problem focus" (p. 16). The

---

[13]More precisely, the committee addressed the identification of "research needs" in its Initial Assessment Report and "problem lists" in its Letter Report. The committee's preference for developing problem lists developed during the course of its later deliberations.

SIDEBAR 3.1 Potential Value of Risk Research to the Cleanup Mission

Risk-related problems deserve special consideration in the science plan, because risk assessments should underpin all attempts to prioritize cleanup activities in the weapons complex. For example, cancer risk from low-level exposure to radioactivity (i.e., near background levels) has never been documented and must be estimated through extrapolation from high exposure data. Basic research on the health effects of low levels of radiation is particularly relevant to the EMSP. At present, the scope or extent of any radiation exposure over the long term is not well defined. Until estimates of the uncertainties are derived, it is not possible to assess realistically the calculated risk estimates.

A genuine need exists to assess systematically and realistically environmental and occupational risk during restoration and post-restoration activities. It is usually assumed, for example, that the environment is protected if the people living in that environment are protected. Risk assessments can help assess the potential for significant changes—adverse or beneficial—in a particular environment due to radiation or chemical exposures. The evaluation of existing data and a determination of the uncertainty associated with each of the many parameters involved in the assessment of end-point risk to a population and the environment can be the basis of a realistic risk assessment. Innovative techniques are necessary to validate quantitative point estimates of the risk over time and to estimate the degree of uncertainty associated with these values.

In the specific area of health assessment, actual health detriments from chemicals are poorly known. The chemicals used at DOE sites are used commercially and have accepted occupational exposure limits. Nonetheless, the scope of the problem for these substances is poorly defined at DOE sites and requires new techniques to evaluate potential exposures.

---

committee returned to this recommendation in its Letter Report: "The committee reaffirms the importance of these summaries and recommends that they be prepared forthwith" (p. 5). An important reason for preparing such summaries is to enable the broader research community, many of whose members are unfamiliar with the DOE cleanup problems, to become sufficiently aware of and interested in these problems so that they may become involved in research of value to the cleanup mission. As the committee wrote in the Initial Assessment Report: "These summaries should be produced for wide circulation to the research community and should be updated as appropriate to reflect current needs" (p. 17). In its Letter Report the committee encouraged the Department to emphasize in these summaries the problems to be solved, rather than the research areas currently viewed as most relevant to their solution.

The committee noted in its Letter Report that it did not have the experience or expertise to provide a list of EM problems that should be addressed through basic research. The committee can, however, provide advice on a process for developing such a problem list, which would form the core of the EMSP science plan. **To this end, the committee recommends both a near-term and a long-term process for developing a science plan for the EMSP. For the near term (i.e., the fiscal year 1997 [FY97] competition), the committee recommends that the Department develop the science plan from existing Department documents.**[14] A description of the EM Science Program prepared by EM, ER, and DOE laboratory representatives in 1995[15] and the previously referenced BEMR report could serve as good starting points for this effort. **For the longer term (i.e., the FY98 competition), the committee recommends that the Department consult with its "problem holders"—the technical staff, managers, and stakeholder advisory groups at the sites who have some understanding of cleanup issues—to obtain guidance on cleanup problems that cannot be addressed practically or efficiently with current knowledge or technologies.** The committee recognizes, of course, that the technical expertise and knowledge for assessing cleanup problems among these groups is uneven and, consequently, suggestions from these groups will

---

[14]Examples of documents that could be used to prepare such summaries include the Baseline Environmental Management Report (see footnote 3 in Chapter 2); DOE, Office of Energy Research, 1990, Basic Research for Environmental Restoration, DOE/ER-0482T (Washington, D.C.: DOE); R. E. Gephart and R. E. Lundgren, 1995, Hanford Tank Cleanup: A Guide to Understanding the Technical Issues, PNL-10773 (Richland, Wash.: Pacific Northwest Laboratory).

[15]DOE, 1995, Description of the Environmental Management Science Program: Working Draft.

have to be considered against that knowledge. Nevertheless, the committee believes that these groups can provide valuable perspectives on the urgency of various cleanup problems at the sites.

Given the large number of DOE sites, these consultations will have to be structured carefully to be manageable by and useful to EMSP staff. For example, each of these three groups (i.e., technical staff, managerial staff, and the site's advisory group) at each of the major DOE sites could be asked to prepare a short (e.g., 5-page) document listing the most important (e.g., in terms of cost and risk reduction) longer-term cleanup problems that cannot be addressed practically with current knowledge or technologies. These papers could then be collected and reviewed by a panel consisting of ER and EM program managers, selected investigators in relevant disciplines, and representatives from the sites. This panel could then generate statements of problems that could be addressed by basic research.

COORDINATING THE INVESTMENT IN BASIC RESEARCH

The science plan developed through the processes described above is likely to be very broad in scope—both in terms of the range of problems and the disciplinary coverage—reflecting the broad scope of the EM cleanup mission and the large number of very difficult cleanup problems across the complex. Indeed, the committee expects that the science plan will require an investment in basic research that is larger than the current $50 million annual investment in the EMSP. To implement the science plan, Department staff should find ways to utilize relevant research being sponsored in other federal programs and to focus the EMSP on those problems that are unique to the weapons complex. At the same time, Department staff also should find ways to inform managers and principal investigators (P.I.s) in these other programs of its needs for research as articulated in its science plan.

Given the relatively small size of the EMSP and its staff, the committee does not deem it prudent to recommend formal coordination mechanisms between the EMSP and other research programs. The committee does, however, offer the following mechanisms as examples of the kinds of coordinating activities that could be of value to the program:

•   *Identify potentially complementary research programs in other federal agencies and provide copies of the science plan to the program directors.* On p. 19 of its Initial Assessment Report, the committee identified examples of programs that sponsor research of relevance to the EMSP. The committee has gathered additional information on federal research programs and provides a more comprehensive list of relevant programs in Table 3.1. Since many of these programs are headquartered in Washington, D.C., staff can be brought together at relatively low cost to participate in meetings where research results are presented and discussed. EMSP staff should consider organizing such a meeting around the release of its science plan for the EMSP.

•   *Obtain and review the reports issued by these programs to become familiar with the P.I.s and research projects.* Many research programs issue annual reports that contain project summaries and publication lists, and some agencies are now beginning to post this information electronically where it can be accessed easily and searched readily.

•   *Invite research program directors from other federal agencies and, when appropriate, investigators supported by their programs to meetings of EMSP investigators and technology users (e.g., the problem holders at the sites), as discussed briefly in Chapter 5.* Such meetings could provide efficient mechanisms to help in applying research results to cleanup and in fostering collaborations between investigators in different disciplines who would not otherwise have a reason to associate.

•   *Attend, where possible, the investigator meetings for other research programs to become familiar with the research projects and P.I.s.* Many research programs bring groups of their P.I.s together periodically to provide progress reports of their work. By carefully targeting these meetings, EMSP staff can become more widely informed of relevant research sponsored by other programs. These meetings also offer opportunities for EMSP staff to alert others about the needs and activities at DOE facilities.

The committee notes that EM and ER staff are already beginning to take proactive steps along these lines. For example, ER has involved an EM staff member in the management group for its NABIR program

(Table 3.1). Additionally, EM staff have initiated contacts with EPA staff to discuss that agency's risk-related research.

As EMSP staff become more knowledgeable about relevant research efforts in other programs, they will be able to move the focus of the EMSP to high-priority problems that are not being addressed elsewhere. At the same time, EMSP staff will be able to identify relevant research from other programs and help move it into technology development efforts. The net effect of these activities is a multifront attack on the science plan and a more effective application of results to the cleanup mission.

BROADENING THE INVESTIGATOR COMMUNITY

The committee's previous reports have made frequent references to broadening the community of investigators involved in the EMSP and to developing a core or "committed cadre" of investigators who are knowledgeable about EM's problems. The committee believes that the Department can take several steps over both the near term and the long term to improve its outreach to the research community and thereby hasten the development of this core group.

The committee noted in its Initial Assessment Report that the long-term success and effectiveness of the EMSP will depend to a large extent on the degree to which the program is able to attract high-quality researchers. In the committee's opinion, EMSP should not be viewed as just another program to support the established environmental research community. Rather, the program should strive to attract creative investigators who do not now work on the Department's problems. This will require significant outreach to the scientific and technical communities, particularly to those not currently engaged in work related to energy research or environmental management. Many of the suggestions offered in the previous section on program coordination will be of benefit to the Department in its efforts to attract "new" investigators to the EMSP.

As noted in the Initial Assessment Report, high-quality researchers can be found in a broad spectrum of the nation's research institutions, including universities, industry, national laboratories, and other federal agencies; investigators in each of these institutions bring unique strengths and perspectives to the program.

TABLE 3.1 Other Federal Research Programs of Relevance to the EMSP

Program Name	Description	Budget
DOD	Supports defense-related fundamental research in physics, chemistry, terrestrial science, ocean science, atmospheric and space science, biological science, materials science, and computer science through the Strategic Environmental Research and Development Program and others.	NA[a]
DOE Energy Research Programs	Supports energy-related fundamental research in bioscience, chemistry, computing, geoscience, health, materials science, and physics through several programs.	NA[a]
DOE-ER Natural and Accelerated Bioreme- diation Research (NABIR) Program	Supports research and development in bioremediation, especially in situ bioremediation of contaminated soils, sediments, and ground water at DOE facilities.	FY 96: $20 M
DOE/EPA/NSF/ONR Joint Program on Bioremediation	Supports bioremediation research with the goal of understanding the factors that impact the risk posed by waste chemicals and their degradation products to ecosystem and human health during the process of bioremediation.	FY 96: $5 M
EPA National Center for Environmental Research and Quality Assurance	Supports research in support of EPA program priorities, including exploratory research, ecosystem indicators, issues in human health risk assessment, endocrine disruptors, ambient air quality, health effects and exposures to particulate matter and associated air pollutants, drinking water, and contaminated sediments.	FY 97: $35 M

NSF Basic Research Programs	Supports fundamental research in bioscience, chemistry, computing, engineering, geoscience, materials science, and physics.	NA[a]
NSF Environmental Geochemistry and Biogeochemistry	Supports interdisciplinary research on chemical processes that determine the behavior and distribution of inorganic and organic materials in the near-surface environment.	FY 97: $5 M
NSF/EPA Partnership for Environmental Research	Supports grants for research in the subjects of water and watersheds, technology for sustainable development, and decision making and valuation for environmental policy.	FY 97: $12 M
USGS Toxic Substances Hydrology Program[b]	Supports USGS research on fate and transport of toxic substances in the nation's hydrologic environment.	FY 97: $14 M

NOTE: Programs are extramural except where indicated. DOD = U.S. Department of Defense, DOE = U.S. Department of Energy, EPA = U.S. Environmental Protection Agency, NA = not available, NSF = National Science Foundation, ONR = Office of Naval Research, USGS = U.S. Geological Survey.
[a]No budget figures are available because environmentally related basic research is not broken out of DOD's, ER's, or NSF's basic research budgets.
[b]Intramural research program.

- *National laboratory investigators:* Many national lab investigators are familiar with the weapons complex and the cleanup mission, and they possess specialized knowledge, facilities and equipment, and analytical and monitoring capabilities. Many of these investigators also are experienced in working in large teams that may be useful to address certain types of multidisciplinary problems.
- *University investigators:* Many university investigators are at the forefront in the fundamental scientific disciplines where advances in knowledge are likely to provide significant future payoffs to the cleanup mission. University investigators also are primarily responsible for training future generations of investigators.

• *Industry investigators:* Like their national laboratory counterparts, many industry investigators have access to specialized knowledge, facilities, and equipment, and many are experienced in working in multidisciplinary team environments at the interface between research and application.

• *Investigators at other federal agencies:* Many federal "mission" agencies have capabilities for addressing problems relevant to EMSP. For example, some agency investigators are involved in work at "testbed" sites[16] on "generic" problems such as ground-water contamination by chlorinated solvents, petroleum hydrocarbon mixtures, and certain heavy metals. Research that utilizes these testbeds can provide new knowledge that can be applied directly to cleanup of the weapons complex.

Over the near term, the Department can broaden the community of investigators concerned with its cleanup problems by encouraging appropriate collaborations among investigators at these institutions. These collaborations are not an end in themselves but rather a route for stimulating new research, introducing new investigators to the Department's problems, and assuring relevance of the projects. Collaborations almost always develop from a perceived need on the part of investigators that additional expertise is necessary to tackle research problems. Thus, the nature of the problems articulated in the science plan may be important for encouraging collaborations in the program.

In particular, collaborations between university investigators—who generally speaking have a great deal of disciplinary expertise but not much knowledge of the Department's cleanup problems—and their national laboratory and industry counterparts can bring a new pool of largely untapped talent to bear on the Department's problems. Additionally, collaborations between investigators and site contractors can facilitate work directly at the sites and ensure its coordination with ongoing cleanup activities. Of course, for this arrangement to work, the contractors may need financial or programmatic incentives, especially when such collaborations result in extra expense, including personnel costs. One way to encourage such collaborations would be to have EM

---

[16]The USGS, EPA, and DOD, among others, operate and/or conduct research at such sites.

program staff arrange for such support through the contractor's cleanup contract.

The committee recommends that collaborations be encouraged where appropriate—but they should not be a requirement for the program. Attempts to force collaborations could discourage some talented scientists from applying to the program. As the committee noted in its Initial Assessment Report, much of the nation's best science continues to be done by single investigators working on individual projects.

Over the longer term, the Department can promote the development of a "committed cadre" by encouraging graduate and postdoctoral training in areas of interest. Such training not only contributes to building a high-quality community of investigators concerned with EM's long-term cleanup problems, but it also brings fresh perspectives and new ideas to bear on the program's problems. **The committee reaffirms the recommendation from its Letter Report (p. 4) that the program "should encourage (but not require) graduate student involvement in research proposals submitted to the program." The committee would add to this recommendation that appropriate postdoctoral training opportunities, including training opportunities within current DOE programs, also should be encouraged to sustain the interest of talented young scientists.**

If the EMSP budget increases in size to the levels indicated in the next chapter, EMSP staff should consider establishing fellowship programs to support highly qualified graduate students, postdoctoral investigators, and early-career scientists. At the graduate level, such fellowship programs would encourage promising students to obtain advanced degrees in academic disciplines relevant to environmental cleanup at DOE. At the postdoctoral and early-career levels, such fellowship programs would steer new Ph.D.s into research careers in fields related to the DOE cleanup mission.

At increased budget levels, the EMSP also could support workshops, seminars, and lectureships to provide an open forum for presentation of results of EMSP-supported research. Seminars at national laboratories and universities by prominent scientists within the EMSP program could be especially helpful in establishing productive collaborations. The topic of workshops and seminars is addressed again in Chapter 5.

4

PROPOSAL SELECTION AND FUNDING

In its Initial Assessment Report, the committee devoted considerable attention to the Department's process for proposal solicitation and selection. The committee's comments in that report focused on the FY96 solicitation and proposal review process, which was well under way when the committee began its work.[1] The committee's Letter Report focused primarily on the content and structure of the FY97 program announcement, but the committee also offered suggestions on the FY97 review process. The purpose of this chapter is to summarize and extend the comments from these previous reports to address the committee's charge (Appendix A) to provide advice on the structure and operation of the program. The comments in this chapter address the following issues:

- review process,
- program funding, and
- the role of stakeholders in the program.

Additional comments on program management can be found in the next chapter.

PROPOSAL REVIEW PROCESS

The Environmental Management Science Program (EMSP) employs a two-stage review process to evaluate proposals submitted to the program—a review of scientific and technical merit followed by a review to assess relevance to the cleanup mission. The merit reviews are performed by panels of scientists and engineers convened by Office of Energy Research (ER) staff, whereas the relevance reviews are performed by panels of Office of Environmental Management (EM) program managers who are familiar with the Department's cleanup

---

[1]As noted in Chapter 1, the FY96 program announcement was published in February 1996, and full proposals were due in May, during the early stages of the committee's study.

problems. This proposal review process has received considerable scrutiny from the committee in its previous reports. In general, the committee has been satisfied with the design of the review process—as noted, for example, in the following excerpt from page 6 of its Letter Report:

> The committee reaffirms its endorsement (from the Initial Assessment Report) of the two-phase review process used in the FY1996 competition that first evaluates the scientific and technical merit of the proposals and then examines more closely the relevance of the proposed work to the clean-up mission. The committee believes that this two-phase review process should continue in FY1997 and that it should continue to be managed as a partnership between ER and EM.

However, this satisfaction is based entirely on the results of the FY96 program competition—which may or may not be typical of future competitions.

As noted in Chapter 1, the committee received extensive written documentation on successful proposals from the Department, including principal investigator (P.I.) and co-P.I. names and affiliations, biographical sketches of P.I.s, abstracts of funded projects, and amounts of other current DOE funding. The committee reviewed these data, and individual committee members paid particular attention to those projects that were within their areas of expertise. Based on this review, the committee reached the following conclusions about the FY96 proposal competition:

• Meritorious projects appear to have been selected in the FY96 proposal competition. This is a qualified judgment, however, because the committee was not able to examine the unsuccessful proposals to determine whether they were qualitatively different from funded proposals. The committee was unable to ascertain what criteria were used in the selection process and, as a consequence, whether these criteria were ones with which it would agree.

• Collaborative efforts were well represented among the list of successful projects. As shown in Table 4.1, about two-thirds of the projects supported in the FY96 competition involved collaborations.

• At least 33 of the 140 P.I.s supported in the FY96 competition currently do not have other Department of Energy (DOE) funding, suggesting that the Department was successful in attracting some "new" researchers to the program.

• The committee was able to obtain firsthand information on the membership of one of the review panels and was able to confirm its overall quality.

The success of this joint review process can be attributed in large part to good communication and coordination between EM and ER staff. In the committee's opinion, a continuing partnership between EM and ER is essential to maintain the effectiveness of the review process.

The committee remains concerned about some elements of the review process, particularly the interaction of the merit and relevance review panels. Basic research, by its very nature, is not usually measured against the yardstick of "relevance." Thus, the relevance review, unless carefully managed, has the potential to compromise the outcome of the merit review process. This could happen if, for example, the relevance review panels were to select many proposals that ranked lower in the merit review instead of more highly ranked proposals. This would have the effect of diminishing the overall quality of the science in the EMSP, which could reduce the long-term effectiveness of the EMSP to the cleanup effort. It also would have the effect of diminishing the influence of merit review panelists on the final outcome of the competition and could discourage highly regarded scientists from serving on EMSP merit review panels.

The committee has two concerns about the transparency and technical credibility of the merit review process, concerns that were expressed in its Letter Report. First, as presently managed, the merit review process is "opaque" to those who submitted proposals to the program, merit review panelists, and the broader research community. The names of the merit review panelists are kept confidential by the Department, so there is no way for P.I.s to evaluate the intrinsic quality

TABLE 4.1 Investigator Collaborations in the FY96 Proposal Competition Based on Data Received from the Office of Science and Technology, U.S. Department of Energy

Type of Collaboration	Number	Percent
Partnerships involving a single university	27	20
Partnerships involving multiple universities	7	5
Partnerships involving a single national laboratory	22	16
Partnerships involving multiple national laboratories	3	3
Partnerships involving universities and national laboratories	31	21
Partnerships involving universities and industry	1	<1
Partnerships involving universities, national laboratories, and industry	1	<1
No partnerships (i.e., single-investigator awards)	47	34
Information not available	1	<1
Total	140	100

of the proposal reviews.[2] Additionally, the merit review panelists were asked to provide individual scores on proposals, but they were not told how their scores were used by ER program managers to make award decisions.

Second, the merit review panels are not constituted as FACA[3] committees. Consequently, the merit review panelists are allowed to discuss and provide individual scores on each proposal, but the panels as a whole are not allowed to reach consensus on individual proposals or to

---

[2]The committee understands that the Department is not required to keep the names confidential, but it has been its practice to do so.

[3]Federal Advisory Committee Act, Public Law 92-463.

provide ER program managers with a ranking of proposals or to make comparative assessments of proposals. Such assessments become especially important when large numbers of proposals are being reviewed, but only a small number of these proposals can be supported— a problem that is likely to get worse in the next few years if funding for the program is not increased.[4] Collectively, the panelists have much greater knowledge on the subjects of the proposals than individual program managers, and it makes good sense to take full advantage of this expertise in the review process. The current process allows ER program managers to operate fairly autonomously with relatively little visibility in the research community for decisions that are being made in the program.

In its Letter Report the committee recommended that ER constitute its merit review panels as FACA committees. In subsequent discussions with ER staff, the committee learned that DOE is prohibited by law from convening FACA committees that are closed to the public. FACA permits agencies to close meetings to the public if sensitive personal or other information is being discussed—as would be the case for proposal reviews. However, the Department's statutory legislation prohibits it from closing any committee meetings, including those constituted under FACA, except for purposes of protecting national security.[5] A 1991 U.S. General Accounting Office (GAO) report,[6] which also recommended that the Department convene its peer review committees under FACA, acknowledged these legal barriers but recommended that the Department seek a change in its legislation to make the use of such committees possible. In its response, which was included at the end of the GAO report, the Department agreed to seek such a change. To the committee's knowledge, however, no change was ever sought.

---

[4]In FY96, DOE received 810 full proposals in the FY96 competition. Based on individual scores from the merit review panelists, DOE program managers grouped these proposals into one of three categories: 77 proposals were rated as "must fund," 111 as "should fund," and 622 as "don't fund." A total of 140 awards were made, including 73 awards to "must fund" proposals and 67 awards to "should fund" proposals.

[5]15 U.S.C. § 776(b) provides the applicable language.

[6]U.S. General Accounting Office, 1991, Peer Review: Compliance with the Privacy Act and Federal Advisory Committee Act, GAO/GGD-91-48, 30 pp. (Washington, D.C.: GAO).

ER staff have asserted that the FACA process would impose a heavy paperwork burden on the Department. The committee does not doubt that FACA will entail some extra paperwork but notes that other federal agencies such as the National Science Foundation (NSF) and the National Institutes of Health (NIH) are able to meet the paperwork requirements routinely.

The committee recommends that the Department examine the entire review process for the EMSP with the goal of increasing its transparency and technical credibility. To this end, the committee recommends that the Department carry through on its stated intention (in its response to the 1991 GAO report) to seek a change in its legislation to allow FACA proposal review panels—and to convene the EMSP merit review panels under FACA once this change is made.

The committee also is concerned with the lack of timely feedback to proposers—both successful and unsuccessful—on the results of the merit and relevance reviews. In discussions with EM and ER staff at its open meetings, the committee learned that in the FY96 proposal competition panelist reviews were not sent to P.I.s unless requested, and these reviews did not always reflect the discussions in the panel meetings.[7] Consequently, some of the reviews were of limited usefulness to P.I.s in understanding why their proposals were declined or how they could be improved. **The committee recommends that in future competitions the proposal reviews be modified to reflect the discussions at the panel meetings and, further, that applicants receive feedback on the content and result of the reviews in a timely fashion.**

PROGRAM FUNDING

The issue of program funding received considerable attention from the committee in its previous reports, as noted in Chapter 1. The committee's Initial Assessment Report provided comments on the program's annual budget, the Department's initial allocation of funding for non-DOE (i.e., university and industry) and DOE (i.e., national

---

[7]These written reviews were prepared by the merit and relevance review panelists before the panel meetings, and they were not updated to reflect any changes that occurred as a result of the panel discussions before being sent to the P.I.s.

laboratory) proposals, and full funding of successful proposals out of current-year funds. The committee recommended that awards in the 1996 program be fully funded up front to ensure that there would be a relatively constant number of new starts in succeeding years of the program.

In its Letter Report the committee returned to the issue of full funding of proposals and also addressed the developing "mortgage" on future-year budgets. This mortgage developed because the Department was unable to fully fund awards to national laboratory investigators but instead had to commit funding from future-year budgets. In the Letter Report the committee presented a financial analysis for the EMSP based on the funding commitments from the FY96 competition. This analysis provided two scenarios for future funding of the EMSP to illustrate the committee's concerns about the future levels of funding for the program given current commitments on future-year program funds.

The *steady-state funding scenario*,[8] which is shown in Table 4.2, was generated using the following set of assumptions:

• Funding of new awards for non-DOE performers (i.e., university, industry, and other nonprofit performers) is continued at the FY96 level of $43 million for three-year grants, and these awards are funded fully in the first year, as was the case for the FY96 proposal competition.

• The ratio of dollars committed each year to awards to non-DOE performers to the dollars committed each year to new awards to national lab performers remains constant at FY96 levels.

• Awards to national lab performers are paid in equal installments over three years.

• Total annual funding for the EMSP is allowed to increase as necessary to satisfy the foregoing assumptions.

As shown in Table 4.2, to maintain funding for new starts at FY96 levels, the total annual funding for the program would almost triple, to $131 million in FY99, before declining to a steady-state value of

---

[8]Referred to as the *unconstrained funding scenario* in the committee's Letter Report.

$112 million in FY2000. This amount is roughly 225 percent of the current annual budget for the program.

The *constrained funding scenario,* which is shown in Table 4.3, was generated using the following set of assumptions:

• Total annual program funding is constrained to FY96 levels of $50 million.

• As in the steady-state funding scenario, the ratio of dollars committed each year to awards to non-DOE performers to the dollars committed to new awards to national laboratory performers remains essentially constant at FY96 levels.

• As in the steady-state funding scenario, awards to national laboratory performers are paid in equal installments over three years. The first installment is paid during the fiscal year in which the awards were made. The two remaining installments are paid in the two succeeding fiscal years. As shown by the scenario in Table 4.3, for example, the $27 million awarded to national laboratories in FY97 would be paid in three equal installments of $9 million in FY97, $9 million in FY98, and $9 million in FY99.

This scenario illustrates the full effects of the mortgage when national laboratory performers receive funding one year at a time and non-DOE performers receive all of their funding up front. As shown in Table 4.3, the mortgage from the FY96 award cycle creates a significant drain on program funds through FY99. In FY97, for example, only $27 million in new funds is available—$18 million to non-DOE performers and $9 million to DOE performers.[9] Indeed, by FY99 only $10 million in new funds is available to non-DOE performers and $6 million in new funds is available to national laboratory performers, about a quarter of the funding available in FY96.

Based on this analysis, the committee draws the following conclusions about funding for the EMSP: (1) the budget for the EMSP will have to increase significantly to maintain a reasonable number of

---

[9]The FY97 program announcement was released just before this report entered review. It indicates that only $20 million in new funding is available, not the $27 million indicated in the calculation shown in Table 4.3.

TABLE 4.2   Hypothetical Funding for the EMSP When Annual Program Funding Is Allowed to Reach a Steady State

Program	Funds Distributed During Fiscal Year (millions of dollars)						
Fiscal Year	1996	1997	1998	1999	2000	2001	2002
Non-DOE performers							
1996[a]	43	0	0	0	0	0	0
1997		43	0	0	0	0	0
1998			43	0	0	0	0
1999				43	0	0	0
2000					43	0	0
2001						43	0
2002							43
National laboratory performers							
1996[a]	4	23	23	19	0	0	0
1997		23	23	23	0	0	0
1998			23	23	23	0	0
1999				23	23	23	0
2000					23	23	23
2001						23	23
2002							23
TOTAL	47	89	112	131	112	112	112

[a]Results from the FY96 proposal competition.

new starts and competitive renewals with a reasonable distribution of funding between DOE and non-DOE performers; or (2) if the budget remains at current levels, both non-DOE and DOE performers could see about a 75 percent drop in funding for new and competitive renewal projects.

In discussions with the committee, EMSP staff have stated that DOE financial practices do not permit them to provide full funding for multiyear proposals from DOE performers. ER staff told the committee that the Director of the Office of Energy Research would like to change these practices and provide full funding for national laboratory proposals

TABLE 4.3   Hypothetical Funding for the EMSP when Annual Program Funding Is Constrained to $50 Million

Program	Funds Distributed During Fiscal Year (millions of dollars)						
Fiscal Year	1996	1997	1998	1999	2000	2001	2002
Non-DOE performers							
1996[a]	43	0	0	0	0	0	0
1997		18	0	0	0	0	0
1998			12	0	0	0	0
1999				10	0	0	0
2000					25	0	0
2001						20	0
2002							17
National laboratory performers							
1996[a]	4	23	23	19	0	0	0
1997		9	9	9	0	0	0
1998			6	6	6	0	0
1999				6	6	6	0
2000					13	13	13
2001						11	11
2002							9
TOTAL	47	50	50	50	50	50	50

[a]Results from the FY96 proposal competition.

in some of its programs but has so far been unable to do so. Indeed, ER staff indicated that they are finding it increasingly difficult to provide multiyear funding for university proposals, even in regular ER programs.

The committee believes that it is beyond its charge to evaluate the Department's current financial practices or to assess the likelihood that these practices can be changed in time to impact the FY97 proposal competition. Nevertheless, the committee continues to be very concerned about the full funding issue because of its potentially significant impacts on future project awards. Simply put, the program must be large enough to support a significant number of "new starts" (i.e., new projects or competitive renewals) each year if it is to be successful in attracting innovative proposals from outstanding researchers who are not now doing research relevant to EM's problems.

The committee believes that, without some assurance that funding will be available to support a reasonable number of new awards annually, EMSP will simply not be viewed as "worth the effort" by potential proposers. Over time this situation is very likely to adversely affect the quality of the program and to diminish its potential benefit to the overall EM program.

The committee notes that DOE itself recognized that EMSP should be a significantly larger program, on the order of $150 million (as expressed by Thomas Grumbly, then-Assistant Secretary for Environmental Management, in the document entitled *Summary of Workshop to Initiate the Development of a Science Program to Support the Department of Energy's Office of Environmental Management*[10]).

The committee appreciates the difficult budget environment that DOE now finds itself in and recognizes that any increases in the budget for the EMSP may be at the expense of other Department programs. In the committee's view, however, this funding should not come from existing ER programs, which are vital to the Department's long-term mission and are an important part of the nation's basic research portfolio. Nevertheless, the EMSP cannot live up to its potential without careful consideration by DOE of both the total funding levels and the funding patterns (i.e., the balance between new and continuing awards). **The committee urges DOE to find a solution to the problem of not being able to "forward fund" projects at national laboratories and reiterates its recommendation from the previous reports to fully fund all awards in the first year.**

ROLE OF STAKEHOLDERS IN
PROPOSAL REVIEW AND SELECTION

During the course of this study, the Department held workshops at three of its sites—Hanford, Savannah River, and Idaho—to inform

---

[10]This workshop was held at the Holiday Inn, Washington Dulles Airport, on July 21, 1995.

stakeholders[11] about the EMSP and obtain feedback on the kinds of cleanup problems that would benefit from basic research. The workshops were attended by DOE staff, contractors, national laboratory and university researchers, members of citizens' advisory groups, and other interested members of the public. The committee did not participate formally in any of these workshops, but two members of the committee and one member of the staff attended two of the workshops as observers. They found the workshops to be useful for providing information to stakeholders about the EMSP and generating some enthusiasm among the stakeholders for this program but less useful for obtaining feedback on research needs.

These workshops were organized because EM staff recognize that stakeholders have legitimate fiscal and programmatic concerns about the EMSP. In particular, stakeholders have an interest in ensuring that the EMSP is using its financial resources—resources that might otherwise be used for cleanup—effectively and that the research sponsored by the program is addressing important problems at the sites.

In the committee's opinion, Department staff have a responsibility to keep the stakeholders informed of this program and to seek their input in defining the site problems for the EMSP science plan. The committee suggested a process in Chapter 3 for obtaining this input. At the same time, some stakeholder groups, particularly industry and government agencies, can assist with the transfer of research results into cleanup. The committee suggests a process for this transfer in Chapter 5.

The committee does not believe that stakeholders should be involved in the day-to-day management of the program, particularly the proposal review and selection process. Proposal review and selection should be based primarily on expert judgments of the intrinsic merit of the proposed research, the feasibility of the technical approach, the competence of the principal investigators to undertake the proposed research, and the adequacy of the facilities for carrying out the proposed work. To be effective and credible, the review and selection process should be carried out by technical experts and should remain free of local concerns and special-interest pressures.

---

[11]A stakeholder is defined by the Department as anyone with an interest in DOE activities or anyone who may be affected by DOE activities. This definition was taken from the EM Primer, which is posted on the Department's Web page.

Having said this, the committee also believes that participation of EMSP investigators in the proposal selection process would be very helpful in future years. As the program matures, these individuals can bring an important perspective that helps link EMSP more closely to the broad research community, which will benefit the process of shaping the longer-term character of EMSP.

DOE should also improve and enhance the ways in which it informs the potential users of EMSP results (e.g., technology managers at the various sites) about the process and the outcome of EMSP proposal selection. In this way the problem holders will become more aware of the kinds of research and the quality of the people that EMSP supports. The hoped-for result of such improved information flow is that these problem holders become more attuned to the long-term benefits of the EMSP to their efforts.

5

MANAGEMENT OF THE EMSP

This chapter addresses the Statement of Task questions related to management needs for the Environmental Management Science Program (EMSP) (Appendix A). The Statement of Task directs the committee to provide advice on evaluation of the basic research supported by the EMSP and its impact on the cleanup mission, as well as the overall structure and management of the program. The committee summarizes the conclusions from its previous reports and provides additional comments in this chapter on the following issues:

- long-term management strategies,
- maintaining program quality,
- assessing outcomes, and
- applying the results of basic research.

LONG-TERM MANAGEMENT STRATEGIES

The conference report that created the EMSP directed that the program be managed by the Office of Energy Research (ER). The Secretary of Energy subsequently decided to establish the program as a joint effort between the Office of Environmental Management (EM) and ER to ensure a continuing focus on both research merit and program relevance. The committee endorsed this joint management approach in its earlier reports and most members of the committee remain convinced that such an approach is necessary for the continued success of the program.

During the short time the program has been in operation, EM and ER staff have worked within a management structure that provides similar levels of responsibility for both offices. Most of the management processes were put into place during the first proposal competition, and many of these processes have yet to be tested through a full project cycle. The fact that EM and ER staff were able to establish this "hybrid" management structure in the middle of a proposal competition attests both to their dedication and energy. It also is a testimony to the efforts made by ER and EM management to devote some of their best people to this program.

During the course of this study, the committee received considerable information from the Department describing the joint management structure for the EMSP.[1] The committee summarizes its understanding of this structure below.

- EM and ER are described as "partnering" at headquarters to set policy for the EMSP and to carry out key tasks such as (1) assuring the quality of ongoing research, (2) determining future research needs, and (3) strengthening the linkage between research and the cleanup activities. Each EMSP project has both a designated ER program manager and an EM program manager.

- Actual administration of EMSP projects is done through the Idaho National Engineering Laboratory (INEL) field office. This office is described as the lead organization to administer, manage, and coordinate the award of research grants. The committee was told that INEL also will be used to pull together information from the focus areas and develop a list of problem needs.[2]

- In the situation where a national laboratory receives a grant, the DOE Operations Office that has oversight for that laboratory also is responsible for administering funding for the award. In addition, the Operations Office coordinates with the headquarters program manager(s) responsible for the award(s) in their laboratory. Some Operations Offices also have the responsibility to identify site-specific needs, to ensure research results are applied, to coordinate interactions with the Site Technology Coordination Groups,[3] to set up various kinds of site-specific workshops, and to do other things that may help with use of the research.

---

[1] The committee received information on the management of the EMSP from several sources, including oral briefings from EM and ER staff at the committee meetings and various written documents prepared in response to committee questions about the program.

[2] Oral presentation from the Director, Office of Science and Risk Policy (DOE-EM), at the committee's seventh meeting, November 18, 1996.

[3] Site Technology Coordination Groups were established at each DOE site to provide prioritized site technology needs lists, to facilitate technology demonstration, and to ensure implementation. The groups also function to inform local regulators of technology development activities and to interact with and solicit input from stakeholders and public-interest groups. The membership of Site Technology Coordination Groups

• The national laboratories also have a role in the program. They are charged with managing EMSP funding for work in their facility; they must put in place mechanisms to promote interactions among inside and outside resources; and they are responsible for organizing and running the topical workshops.

These management activities seem reasonable to the committee when considered individually. In the aggregate, however, these activities and the structure that supports them seem unnecessarily complicated. Indeed, when considered against the small size of the EMSP and its focus on basic research, the management structure seems overloaded with administrators and coordinators.

As the program settles into a "steady state" over the next several years, the committee believes that simplification of program management and a clearer delineation of responsibilities among all management participants is needed to ensure its continued effectiveness. The committee believes that in the future program management staff will take on new, important, and potentially conflicting management responsibilities, for example:

• Maintaining internal and external advocacy for the program.
• Developing and maintaining performance measures for accountability to Congress and stakeholders.
• Developing outreach initiatives to ensure the continuing quality of grant recipients.
• Ensuring the continuing cooperation and coordination between EM and ER.
• Ensuring that the results of the research are utilized at the earliest possible time.
• Enhancing the productivity of the program.

In the committee's opinion, EMSP staff will have great difficulty in executing these tasks effectively under the current management structure where no single individual is "in charge" of the program. **Therefore, the committee recommends that management of the EMSP be vested in a**

---

includes personnel from the site's operations offices, contractor and national laboratory personnel, and EM Program personnel.

single individual—an EMSP Program Director—who should have authority, responsibility, and accountability for meeting the program's objectives.

The EMSP Program Director should be an individual with a research background who is respected within the research community and understands the mission and responsibilities of EM. The Program Director must have access to and be included in the strategic planning activities within EM and must be utilized by EM management as an important scientific voice in the planning of the EM research and technology development agenda. Involvement in the latter activity is particularly important to achieve the earliest deployment of EMSP research results into technology development and, ultimately, cleanup activities. Similarly, the Program Director must be included in the planning activities within ER to ensure the proper coordination of the EMSP with other ER research programs. The Program Director also should be responsible for ensuring outreach and coordination activities among performers and stakeholders. He or she must have the responsibility to set policy for grant administration and reporting requirements and provide direction to the program managers who oversee the proposals and grants on a day-to-day basis.

The Program Director must have the support of both the Director of the Office of Energy Research and the Assistant Secretary for Environmental Management to utilize the considerable resources of both organizations for the benefit of the EMSP. At the same time, the Program Director must be able to balance the interests of ER (to support high-quality basic research) and EM (to support research that is relevant to the cleanup mission) and must have the authority to resolve conflicts when these interests come into competition. In the committee's view, the Program Director can be effective in achieving and sustaining this balance only if she or he is functionally independent of both EM and ER. **To allow for such independence, the committee recommends that the EMSP Program Director report to the Under Secretary for Energy.**

The committee spent a great deal of time discussing alternative management strategies for the EMSP before making the recommendations that appear above. In fact, the committee considered the following five alternatives: (1) status quo, that is, joint management by EM and ER with no Program Director, (2) management of the program by ER with management responsibility vested in an ER Program

Director, (3) management of the program by EM with management responsibility vested in an EM Program Director, (4) joint management by EM and ER with responsibility vested in a Program Director reporting to both EM and ER, and (5) joint management by EM and ER with responsibility vested in a Program Director reporting to another office in DOE.

Initially, the status quo alternative had considerable appeal for the committee. The hybrid organization comprised of EM and ER staff is already in place and has worked well to date. As noted at the beginning of this section, however, the committee concluded that the current arrangement structure would not be workable for addressing the longer-range needs of the EMSP or for balancing the near-term and long-term pressures on the program.

One could interpret the congressional language that established the EMSP (Chapter 1) as supporting ER management with an ER Program Director (alternative 2). This alternative does have appeal—the EMSP is a basic research program, and ER is highly skilled at managing basic research. In fact, ER is now managing the merit review process in the EMSP precisely because this is something it does well.

The major disadvantage of this alternative is that the EMSP would likely lose its strong linkages to the users and their problems, which is not what the Congress intended to happen: "This funding is to be used to stimulate the required basic research, development and demonstration efforts to seek new and innovative cleanup methods to replace current conventional approaches . . ." (H.R. 1905; see Chapter 1). A majority of the committee was of the strong view that alternative 2 would not accomplish this linkage. However, one member of the committee believes that the only way the EMSP program can hope to be successful is if it is managed by ER alone (see Appendix D).

Alternative 3 (EM management with an EM Program Director) has some appeal as well. Under EM management, the EMSP would maintain a strong focus on the cleanup mission and would be more responsive to immediate and site-specific technology needs. However, management by EM would likely result in a shift of emphasis toward projects with more immediate payoffs at the expense of longer-term, higher-risk, or more innovative projects. In addition, EM has little experience in managing basic research programs and peer review of basic research, and it has relatively little contact with the basic research community in universities, national laboratories, or industry. Thus, EM would be on a steep and rocky "learning curve" were it given the management responsibility for this program.

The committee has an additional concern with alternatives 2 and 3: if the EMSP Program Director reported either to EM or to ER, he or she would be driven by the interests of those offices and would find it difficult to operate independently of those interests. The committee believes that it is essential to keep both EM and ER involved in the EMSP because each plays different, largely complementary, and equally important roles in the program.

Joint management with the Program Director reporting to both EM and ER (alternative 4) also was judged to be unworkable by the committee, largely for the same reasons that the current management arrangement was deemed to be unworkable over the long term—namely, the Program Director would likely find it difficult to please two masters having fundamentally different missions.

Thus, the committee settled on alternative 5 (joint management with the Program Director reporting to another office within DOE) largely by a process of elimination. This alternative maintains the productive collaborations that are occurring currently between EM and ER, it gives both offices some "ownership" of the program, and it provides leadership to deal with the longer-term issues identified at the beginning of this section. Further, it puts a single individual in charge of and accountable for the program and allows this individual to balance the competing interests of EM and ER.

The committee recommended that the Program Director report to the Under Secretary because both EM and ER report directly to the Under Secretary's office. The committee recognizes that the recommendation on reporting responsibilities for the Program Director could be viewed as unrealistic when the small size of this program is considered against the other responsibilities of the Under Secretary. Nevertheless, the committee makes this recommendation because it believes that, although small, the EMSP can contribute significantly to the Department's ability to resolve the contamination legacy and to utilize effectively the several hundred billion dollars that has been estimated will be spent on the cleanup effort.

The committee notes that agencies such as the National Science Foundation (NSF), the National Institutes of Health (NIH), and the

National Aeronautics and Space Administration (NASA) have small offices and programs that report to high levels (frequently to the agency heads) within their organizations, particularly when the activities of the offices cross internal organizational lines. These arrangements are frequently transient and are intended to bring visibility, emphasis, coordination or management attention to specific initiatives. Examples of such arrangements include NSF's Office of Science and Technology Infrastructure, NASA's Office of Safety and Mission Assurance, and NIH's Office of AIDS Research and Office of Behavioral and Social Science Research.

The committee understands and appreciates the difficult task the Department of Energy (DOE) faces in creating a basic research program that will serve the needs of a highly goal-oriented organization such as EM. In some respects the committee is troubled by the prospect of a program enmeshed in an irreconcilable conflict between the character of its basic research and the need for this research to be ultimately useful to EM and the cleanup effort, particularly given the small size of the program in relation to the total EM effort. Indeed, there were some committee members who believed that basic research was fundamentally incompatible with the strongly needs-driven mission of the EMSP. The committee discussed various ways that this conflict might be addressed (e.g., setting aside money for "blue sky" projects with no particular relevance in order to reinforce the basic character of the program). But at this early stage and given the program's small size, it is not clear to the committee what the near-term versus long-term pressures will be, so the committee thought it inappropriate to be overly prescriptive because it did not want to drive the program in unproductive directions.

The committee notes, however, that there are other agencies, for example, the NIH, the Department of Agriculture, and the National Oceanic and Atmospheric Administration, where mission-oriented basic research is performed, supported, and managed reasonably well and where the long-term outcome has been both high-quality research and significant advances in achieving those agencies' missions. The issue of how to manage such an effort within DOE was one with which the committee struggled mightily, perhaps in large part because there is still a significant lack of clarity about what EM's mission really is.

MAINTAINING PROGRAM QUALITY

Many federal agencies have found that over time their research programs are strengthened and their credibility reaffirmed through periodic, rigorous, independent peer review of all aspects of the programs. For example, the National Institute of Standards and Technology has for many years used "visiting committees" to review each of its major divisions. These committees are usually comprised of eminent scientists and engineers from industry and academia and often include senior industry managers. NSF also uses such visiting committees in many of its research programs. These committees review the operations of the program or division (i.e., effectiveness of peer review, processing time for grants), and also the program's or division's strategic directions and scientific focus. Many universities also use visiting committees to review the quality of their academic programs—in some cases, members of the committee include representative sponsors of research on campus and can effectively articulate the viewpoint of a "customer."

The committee believes that the EMSP would similarly benefit from periodic, independent peer reviews. These reviews should address all aspects of EMSP program management, including

- the merit and relevance review processes,
- quality of funded proposals,
- effectiveness of the application of research results to technology development and cleanup,
- effectiveness of the program in attracting outstanding researchers and innovative research ideas, and
- overall management efficiency and effectiveness.

The committee recommends that the Department convene an independent review panel at appropriate intervals to review the performance and effectiveness of the EMSP.[4]

---

[4]One of many possible ways to obtain this review is through the existing Science Advisory Panel of the Environmental Management Advisory Board. This panel, which is chaired by Dr. Frank Parker of Vanderbilt University, is charged with examining and evaluating the short-term as well as the longer-term impacts of the EMSP program on the cleanup effort.

ASSESSING OUTCOMES

The committee recognizes that the *long-term* success of the EMSP depends on the quantity and quality of the "outcomes"—namely, the impacts on fundamental scientific understanding and, ultimately, on cleanup. However, the time scale for basic research may be quite long. The committee also recognizes that the measurement of outcomes from basic research is currently receiving thorough and careful consideration by many federal research agencies.[5]

At present, no criteria have been established to measure outcomes from the EMSP, although EM staff have proposed two performance criteria to provide such measurements: (1) the number of research projects tied to science needs as identified by Site Technology Coordination Groups, site-specific science research agendas, and program offices and (2) the number of research projects with documented peer-reviewed research results.[6]

In view of the wide breadth of disciplines supported within EMSP and the well-recognized problems of assessing performance of basic research,[7] the committee advises the Department against attempting the development of a general, formal quantitative structure for assessing the performance of the work of its investigators. Nevertheless, the committee believes that it will be essential to review and assess the quality of EMSP on a periodic basis. In the committee's view, the most

---

[5]National Academy of Sciences, National Academy of Engineering, Institute of Medicine, 1996, An Assessment of the National Science Foundation's Science and Technology Centers Program (Washington, D.C.: National Academy Press).

[6]Presentation to the committee by the Associate Deputy Assistant Secretary for Science and Risk Policy at the committee's fifth meeting, September 27, 1996.

[7]See, for example, National Research Council, 1994, Quantitative Assessments of the Physical and Mathematical Sciences: A Summary of Lessons Learned (Washington, D.C., National Academy Press); National Research Council, 1995, Research Restructuring and Assessment: Can We Apply the Corporate Experience to Government Agencies?, (Washington, D.C., National Academy Press); National Research Council, 1995, Allocating Federal Funds for Science and Technology (Washington, D.C., National Academy Press); Office of Technology Assessment, 1986, Research Funding as an Investment: Can We Measure the Returns? OTA-TM-SET-36 (Washington, D.C.: Office of Technology Assessment); R. N. Kostoff, 1993, Semiquantitative methods for research impact assessment, Technological Forecasting and Social Change 44(Nov.):3; National Science and Technology Council, 1996, Assessing Fundamental Science (Washington, D.C.: Office of Science and Technology Policy).

important component of an evaluation of research performance is through a review of the work of investigators supported by the program by an independent review panel of leaders in the field. Such review will assess the overall scientific quality of the program and the extent to which the research it supports has led to technical or intellectual "breakthroughs" of value to the scientific community and technology development efforts.

Despite the acknowledged limitations of review by peers, no better means has been found to evaluate and assure research quality over the long term. As noted in a recent report by the Office of Science and Technology Policy,[8] "for evaluating current programs in individual agencies, merit review based on peer evaluation will continue to be the primary vehicle for assessing the excellence and conduct of science at the cutting edge." Ultimately, of course, it will be the quality of the panel members carrying out such reviews that will determine the quality of the EMSP-supported research.

The committee recommends that the independent review panel be charged with the responsibility of assessing the quality of EMSP science and its impacts.

To accomplish this task, the panel should be provided with information about EMSP by the Program Director that includes but is not limited to the following:

- a comprehensive listing of publications by EMSP grantees;
- a listing of graduate and postdoctoral students trained by EMSP investigators; the degrees, if any, awarded; and current positions of these students;
- a compilation of the most significant scientific results of EMSP with a discussion of how these were selected;
- a compilation of the linkages to the larger EM effort developed with EMSP-supported research; and
- where possible, retrospective studies of the long-term impacts of EMSP results on technology development and cleanup.

The committee recognizes that it could take several years for the compilation of this information to be meaningful even for an initial assessment of the quality of science and its impact. This is inherent in the

---

[8]National Science and Technology Council, 1996, Assessing Fundamental Science (Washington, D.C.: Office of Science and Technology Policy).

nature of a basic research program. But the committee also recognizes that there are shorter-term "drivers" for program assessment, notably the Government Performance and Results Act of 1993. The Department must provide information to the Congress on an annual basis about its performance in response to the mandate of GPRA. Indeed, there may be some advantages to the Department and the EMSP in considering performance measures that have somewhat more immediacy than those listed above but that recognize that the "payback time" for EMSP as a basic research program will be long. Such shorter-term measures might help to sustain the interest in and commitment to EMSP of managers of technology development and cleanup activities. For example, such assessments might focus on processes for evaluating the quality of research proposals and for applying results to cleanup. This might then help to reinforce the linkages to the larger EM effort, the impact of which could only be fully assessed years later.

The committee, therefore, recommends that the Program Director assume the responsibility for developing a "portfolio" of information that would support both short-term and long-term assessment of EMSP by the independent review panel. The Program Director might be well served in this regard by exploring what strategies are being used by other federal agencies that support basic research.

However, the committee believes that attempts at short-term assessments of basic research programs such as EMSP will have very limited value at best. Information of the kind noted above, namely long-term data on outcomes and impacts, is, indeed, the most effective way to assess the value of EMSP and presents the most complete picture of both the quality of the EMSP research activities and their ultimate impact on the cleanup mission.

APPLYING THE RESULTS OF
BASIC RESEARCH TO CLEANUP

The EMSP is designed to support high-quality basic research that has the potential to have significant positive impacts on the broader cleanup effort. It is not possible to predict when and where such impacts will occur. What can be predicted is that by supporting high-quality basic research, new knowledge and insights will be gained and, over time, the benefits from such knowledge and insights will pay off in the broader cleanup mission.

The movement of new knowledge and insights from investigators to full-scale application is a slow and diffuse process—a process without clear pathways in most cases. **As a way of facilitating this information flow and stimulating new research ideas, the EMSP Program Director should convene annual workshops, seminars, and symposia that bring together EMSP investigators, program managers from EM and ER (including those in the EM focus areas), site contractors and other "problem holders," and, when appropriate, other stakeholders, regulators, and principal investigators (P.I.s) and managers from other research programs. The Program Director should assume responsibility for determining how to best structure such activities so that they serve the interests of investigators and EM's needs for information transfer.** Of course, such gatherings should not take the place of papers and reports, which, particularly when peer reviewed, form the basis for wide communication among scientists. However, the committee cautions that whatever mechanisms are developed, they must add value to the EMSP and should not be simply a check mark on a "to do" list.

It will be important in any effort that is undertaken to improve communication and information flow to involve the problem holders at the sites. These individuals will not only have the greatest knowledge about the sites but will also be able to assist in integrating the results of EMSP into the long-term EM effort. The ultimate success of EMSP may depend in no small part on the support and participation of these problem holders.

The responsibility for disseminating results from EMSP is not EMSP's alone. Other offices in EM, especially the other parts of the Office of Science and Technology, must take an active role in ensuring that the Department and the nation reap the full benefits from EMSP-supported research. It is beyond the committee's charge to advise the Office of Science and Technology on how to move the research from the EMSP and other federal research programs into application—the committee simply notes that, without an active effort to move research into technology development and application, the EMSP will become a high-quality research program with a limited impact on EM's cleanup mission.

APPENDIXES

APPENDIX A

STATEMENT OF TASK

The committee will produce two reports that address the science and management needs of the Department of Energy's (DOE's) Environmental Management (EM) Science Program. These reports will be produced in two separate activities as noted below.

ACTIVITY #1: FY97 RESEARCH PROGRAM

The committee will draw on the expertise of its members and other outside experts, the results of the 1996 DOE workshops on research needs, and previous National Research Council (NRC) and federal government reports in order to address the following questions:

1. How can basic research be used to help DOE EM "to complete its mission successfully in the next few decades"?
2. How can a basic research program help add value to DOE EM's cleanup efforts?
3. What kinds of technical challenges would likely benefit from a program in basic research?
4. How can the research program take advantage of the unique capabilities of U.S. universities and federal labs?
5. How can the research program take advantage of research efforts and capabilities in other DOE programs and other federal agencies?
6. What, if any, additional areas of research should be included in the fiscal year (FY) 1997 program announcement as the DOE EM Science Program evolves?

The committee will not attempt to be comprehensive in addressing these questions, but, rather, its focus will be on providing guidance to DOE-EM for use in the FY97 program solicitation.

ACTIVITY #2: SCIENCE AND MANAGEMENT NEEDS

The committee will produce a final report that provides a more detailed assessment of the science and management needs of the EM Science Program. This report will address the following questions:

Science Needs

1. How can science needs most effectively feed into the development of the EM research agenda?

2. How can the research program be structured to take advantage of research efforts and capabilities in other DOE programs and other federal agencies? (The committee would revisit the issue from the first activity.)

3. How can the research program be structured to broaden the community of researchers that can be called upon to address environmental problems?

4. What areas of basic research are likely to provide the best payoffs for EM cleanup efforts over the next few decades?

5. What additional areas of research should be included in future program announcements as the DOE EM Science Program evolves? (The committee would revisit the issue from the first activity.)

Management Needs

1. How can the DOE evaluate the quality of the basic research it supports and the impact of this research on its cleanup mission?

2. How can DOE identify changing needs for basic research as the program evolves?

3. How should the program be structured and operated in order to assist the DOE in overall reduction of cleanup costs, risks, waste generation, and time requirements?

4. How can the program be structured to take advantage of the unique capabilities of U.S. universities and federal labs? (The committee would revisit the issue from the first activity.)

APPENDIX B

LIST OF PRESENTATIONS

Environmental Management Science Program: Background and History, Carol Henry (DOE-EM), May 11, 1996.

Environmental Management Science Program: Current Process, Michelle Broido (DOE-ER), May 11, 1996.

Panel Discussion on EM Science Program/Opportunities and Challenges, Sally Benson (Lawrence Berkeley National Laboratory), Gregory Choppin (Florida State University), Donald DePaolo (University of California), A. J. Francis (Brookhaven National Laboratory), Remy Hennet (S.S. Papadopulos & Associates), Terry Surles (Argonne National Laboratory), May 11, 1996.

Reflections on the First Committee Meeting, Carol Henry (DOE-EM) and Ari Patrinos (DOE-ER), June 15, 1996.

EM Science: Challenges and Opportunities, Judy Bostock (DOE-Savannah River), June 15, 1996.

Reflections on the First Report, Mark Gilbertson (DOE-EM) and Ari Patrinos (DOE-ER), July 22, 1996.

FY 1996 Proposal Competition: Initial Assessment, Carol Henry (DOE-EM) and Ari Patrinos (DOE-ER), July 22, 1996.

Short Report on Cleanup Challenges at the Hanford Site, Steve Blush (independent), July 22, 1996.

Briefings on Related Research Programs, Cliff Dahm (NSF), Jay Grimes (DOE), and Dorothy Patton (EPA), July 22, 1996.

Selected Views on Cleanup Challenges and Research Needs at the Hanford Site, Billy Shipp and Roy Gephart (Pacific Northwest National Laboratory), Frank Parker (Vanderbilt University), and Deborah Trader (Richland-DOE), July 23, 1996.

Selected Views on Cleanup Challenges and Research Needs at the Savannah River Site, Lou Papouchado (Savannah River Technology Center), Joe Rossabi (Savannah River Technology Center), and Jim Brown (Savannah River Site), July 23, 1996.

Selected Views on Cleanup Challenges at the Rocky Flats Site, Richard Bateman (Kaiser Hill) and Joyce Schroeder (Los Alamos National Laboratory), July 23, 1996.

Briefings on Research Management at National Laboratories, Philip Thullen (Lawrence Livermore National Laboratory) and Thomas Dunning (Pacific Northwest National Laboratory), July 23, 1996.

Briefings on Corporate R&D Management, Margaret Gruzca (Industrial Research Institute), July 23, 1996.

Building Partnerships with Government, Universities, and Industry, Thomas Moss (National Research Council), July 23, 1996.

Briefing on the Lab Coordinating Council, William Schertz (DOE), July 23, 1996.

Briefings on Federal Research Programs, Constance Atwell (National Institutes of Health) and Ronald Kostoff (Office of Naval Research), July 23 and 24, 1996.

Selected Views on Cleanup Challenges and Research Needs at the Idaho Site, John Beller (Idaho National Engineering Laboratory) and Tom Williams (DOE-Idaho), July 24, 1996.

Selected Views on Cleanup Challenges and Research Needs at the Oak Ridge Site, Sharon Robinson (Oak Ridge National Laboratory), July 24, 1996.

GPRA (Government Performance and Results Act), Jack Fellows (Office of Management and Budget), July 24, 1996.

Briefings on Department of Defense Research Programs, Jeff Marqusee (DOE), July 24, 1996.

Lessons Learned from FY96 Proposal Competition; Plans for FY97 Program Announcement; and Expectations for the Letter Report, Steve Domotor (DOE-EM), Mark Gilbertson (DOE-EM), and Bill Luth (DOE-ER), August 21, 1996.

Selected Views of Research Needs from Focus Groups: Landfills and Plumes, Brian Looney (Plumes/Landfills Focus Area, Savannah River Site); *Tanks*, Rod Quinn and John LaFemina (Tanks Focus Area, Pacific Northwest National Laboratory); *Mixed Waste*, John Kolts (Mixed Waste Focus Area, Idaho National Engineering Laboratory); *Decontamination and Decommissioning*, Steve Bossart (Decontamination and Decommissioning Focus Area, DOE-Morgantown Energy Technology Center); August 22, 1996.

Presentations and Discussions on Program Management, Chris Parkinson (PA Consulting Group), Al Sattelberger (Los Alamos National Laboratory), Steve Domotor (DOE-EM), August 22, 1996.

Committee-DOE Discussions on the Following Issues: Reflections on the Last Committee Meeting, Program Management Plans for the EMSP in FY 1997, FY 1997 Financial Plan for the EMSP, Current Science and Technology Integration Efforts in EM-50, Plans for Assessing the Impact of the EMSP on Technology Development and Cleanup, Coordination of EMSP with ER Programs, Full Funding for National Laboratory Proposals; Carol Henry and Mark Gilbertson (DOE-EM), Michelle Broido (DOE-ER); September 27, 1996.

Management of R&D to Application: Experiences from EPRI (Electric Power Research Institute) and GRI (Gas Research Institute), Bob Bell (Consolidated Edison), September 27, 1996.

Management of Mission-Directed Basic Research: Two Examples from ORNL, Janet Cushman (Oak Ridge National Laboratory) and Stan Auerbach (retired, Oak Ridge National Laboratory), September 27, 1996.

Committee-DOE Discussions on the Following Topics: Reflections on the Letter Report, Issues to Be Addressed in the Final Report: Program Management, Financial Plan, Integration of Science into Technology Development, Assessing the Effectiveness of the EMSP, Coordination with Other Research Programs; Carol Henry (DOE-EM), Ari Patrinos, Jean Morrow, and Bill Millman (DOE-ER), October 22, 1996.

Comments from EM and ER on the Final Report, Mark Gilbertson (DOE-EM) and Roland Hirsch (DOE-ER), November 18, 1996.

APPENDIX C

BIOGRAPHICAL SKETCHES OF COMMITTEE MEMBERS AND CONSULTANTS

AHEARNE, John F.—Dr. Ahearne received his B.S. and M.S. degrees from Cornell University and his Ph.D. in plasma physics from Princeton University. He has served as commissioner and chairman of the U.S. Nuclear Regulatory Commission, system analyst for the White House Energy Office, Deputy Assistant Secretary for Energy, and Principal Deputy Assistant Secretary for Defense. He currently is the director of the Sigma Xi Center for Sigma Xi, The Scientific Research Society, and a lecturer in public policy and adjunct professor of civil and environmental engineering at Duke University. Dr. Ahearne is a member of the Department of Energy's Environmental Management Advisory Board and the National Research Council's Board on Radioactive Waste Management and has served on a number of National Research Council committees examining issues in risk assessment. His professional interests are reactor safety, energy issues, resource allocation, and public policy management. He is a fellow of the American Physical Society, American Association for the Advancement of Science, and American Academy of Arts and Sciences and a research fellow at Resources for the Future. He is a member of Sigma Xi, the Society for Risk Analysis, the American Nuclear Society, and the National Academy of Engineering.

ARNETT, Edward M.—Dr. Arnett earned a B.A., M.S., and Ph.D. in chemistry from the University of Pennsylvania. He is R.J. Reynolds Professor Emeritus of chemistry at Duke University and has held prior professorships at the University of Pittsburgh and Western Maryland College. His expertise is in organic and physical organic chemistry. He is a Guggenheim fellow and has received numerous awards, including most recently the Arthur C. Cope Scholar Award and the American Institute of Chemists Distinguished North Carolina Chemist Award. Dr. Arnett is a member of the National Academy of Sciences.

AUERBACH, Stanley I.—Dr. Auerbach earned his B.S. and M.S. from the University of Illinois, and his Ph.D. in zoology from Northwestern University. Dr. Auerbach retired as director of the Environmental Sciences Division at Oak Ridge National Laboratory in 1990. His research interests include radiation ecology ecosystem analysis and radioactive waste cycling in terrestrial ecosystems. Dr. Auerbach's former academic positions include lecturer and adjunct professor at the University of Tennessee and visiting professor at the University of Georgia. He has served on or chaired several National Research Council committees, boards, and commissions since 1961. He is a member of the American Institute for Biological Science, American Association for the Advancement of Science, Ecological Society of America, British Ecological Society, International Union of Radioecologists, and Health Physics Society.

BOUWER, Edward J.—Dr. Bouwer received his B.S.C.E. from Arizona State University in civil engineering and his M.S. and Ph.D. in environmental engineering and science from Stanford University. He is currently a professor of environmental engineering at Johns Hopkins University. His research interests include biodegradation of hazardous organic chemicals in the subsurface, biofilm kinetics, water and waste treatment processes, and transport and fate of bacteria in porous media. He serves on the board of directors for the Association of Environmental Engineering Professors and on the editorial boards for the *Journal of Contaminant Hydrology* and *Biodegradation*. He has served on three past National Research Council committees.

BRAUMAN, John I.—Dr. Brauman earned a B.S. from the Massachusetts Institute of Technology and a Ph.D. in chemistry from the University of California at Berkeley. Dr. Brauman is the J.G. Jackson–C.J. Wood Professor of Chemistry at Stanford University. He began his career at Stanford University in 1963 as an assistant professor. His research interests include physical and organic chemistry, gas-phase ionic reactions, electron photodetachment spectroscopy, and reaction mechanisms. He is the recipient of many awards from the American Chemical Society, including the Award in Pure Chemistry, the James

Flack Norris Award in Physical Organic Chemistry, and the Arthur C. Cope Scholar Award. Dr. Brauman is a Guggenheim fellow and an honorary fellow of the California Academy of Sciences; he is a member of the National Academy of Sciences, the American Academy of Arts and Sciences, and the American Chemical Society. He has served on several National Research Council committees.

HARLEY, Naomi H.—Dr. Harley holds a B.E. in electrical engineering from the Cooper Union and an APC in management from the New York University Graduate Business School. She received an M.E. in nuclear engineering and a Ph.D. in radiological physics from New York University. Dr. Harley is a research professor of environmental medicine at the New York University School of Medicine, where she also serves on the Medical Isotopes Committee. Her expertise is in radiation carcinogenesis, and her major research interests include measurement of inhaled or ingested radionuclides, modeling of their fate in the human body, and calculation of the detailed radiation dose to cells specific to carcinogenesis. She is a member of the National Council on Radiation Protection and Measurements and an adviser to the U.S. Delegation of the United Nations Committee on the Effects of Atomic Radiation. Dr. Harley is a member of the editorial board of *Environment International* and a fellow of the Health Physics Society; she holds three patents at New York University for radiation detection devices. Dr. Harley has published over 100 journal articles.

LEWIS, Harold W.—Dr. Lewis received his Ph.D. in physics from the University of California at Berkeley. He is professor emeritus of physics at the University of California at Santa Barbara, is past director of its Quantum Institute, and specializes in theoretical physics. He has served on the Defense Science Board and has served on and chaired several national committees relating to nuclear safety. These include the American Physical Society study on light-water reactor safety, the Risk-Assessment Review Group of the Nuclear Regulatory Commission, the Advisory Committee on Nuclear Facility Safety of the Department of Energy, and the President's Nuclear Safety Oversight Committee. He also has served on several National Research Council committees. He is the author of *Technological Risk* (New York: Norton, 1992).

LOVLEY, Derek R.—Dr. Lovley received a B.A. in biological sciences from the University of Connecticut, an M.A. from Clark University, and a Ph.D. in microbiology from Michigan State University. He is a professor of microbiology at the University of Massachusetts, Amherst. His research interests comprise the physiology and ecology of novel anaerobic microorganisms, molecular analysis of anaerobic microbial communities, and bioremediation of metal and organic contamination. He is an associate editor for *Anaerobe* and is on the editorial boards of *Applied and Environmental Microbiology*, *Microbial Ecology*, and *FEMS Microbiology Ecology*.

MacLACHLAN, Alexander—Dr. MacLachlan received his B.S. in chemistry from Tufts University and his Ph.D. in physical organic chemistry with a minor in chemical engineering from the Massachusetts Institute of Technology. Dr. MacLachlan is a retired Under Secretary for R&D Management at the U.S. Department of Energy. Prior to his work at the Department, he retired from a long career at DuPont as senior vice president for research and development and chief technical officer. Dr. MacLachlan is a member of the National Academy of Engineering and Phi Beta Kappa. He serves on the Secretary of Energy's External Advisory Board and the Sandia President's Advisory Council at Sandia National Laboratory.

MANNELLA, Gene G.—Dr. Mannella earned a B.S. from Case Institute of Technology and a Ph.D. in chemical engineering from Rensselaer Polytechnic Institute. He retired in 1994 as senior vice president of business operations at the Gas Research Institute, headquartered in Chicago. He has also served as director of the Washington office of the Electric Power Research Institute; vice-president and general manager of Mechanical Technology, Inc.; and senior vice-president at the Institute of Gas Technology. Dr. Mannella has held several positions at government agencies, including the National

Aeronautics and Space Administration, Department of Transportation, and Energy Research and Development Administration (predecessor to the Department of Energy). He has authored numerous technical papers and served on several committees and boards, including the Washington Coal Club.

NOONAN, Norine E.—Dr. Noonan received her B.A. from the University of Vermont, summa cum laude, in zoology/chemistry, and her M.A. and Ph.D. degrees in cell biology and biochemistry from Princeton University. She is vice president for research and dean of the Graduate School at the Florida Institute of Technology in Melbourne. Prior to joining Florida Tech in October 1992, Dr. Noonan was chief of the Science and Space Programs Branch of the Energy and Science Division, Office of Management and Budget. In this capacity she was responsible for legislative programs and combined budgets. Before becoming branch chief, Dr. Noonan was senior budget and program analyst for the branch for four years. She was an American Chemical Society Congressional Science Fellow for the U.S. Senate Committee on Commerce, Science, and Transportation; a research associate professor of biochemistry at Georgetown University School of Medicine; an expert consultant for the Subcommittee on Science Research and Technology; and an associate professor of physiological sciences at the University of Florida, College of Veterinary Medicine. Dr. Noonan is a member and fellow of the American Association for the Advancement of Science and also a member of the American Society for Cell Biology, Sigma Xi, and Phi Beta Kappa.

SACKS, Jerome—Dr. Sacks received his B.A. and Ph.D. in mathematics from Cornell University. He is director of the National Institute of Statistical Sciences, located in Research Triangle Park, North Carolina, and a professor at the Institute of Statistics and Decision Sciences, Duke University. In addition to his previous academic career, Dr. Sacks served as a program officer at the National Science Foundation. He has led an extensive research program in environmental statistics and served on boards and committees of the National Research

Council and its Commission for Physical Sciences, Mathematics and Applications.

SATTELBERGER, Alfred P.—Dr. Sattelberger earned his B.A. in chemistry from Rutgers College and his Ph.D. in inorganic chemistry from Indiana University. He began his research career at the University of Michigan in 1977 and moved to Los Alamos National Laboratory in 1984, where he is currently the director of science and technology base programs. This office has responsibility for internal R&D funding, science education, and university outreach. Dr. Sattelberger's research interests include actinide science, technetium coordination and organometallic chemistry, and metal-metal multiple bonding. He is a member of the Executive Committee of the Inorganic Chemistry Division of the American Chemical Society and serves on the board of directors of the Inorganic Synthesis Corporation and on the editorial board of *Inorganic Chemistry*. He served as a reviewer on the FY96 General Inorganic Chemistry EMSP merit review panel.

SILVER, Leon T.—Dr. Silver earned a B.S. in civil engineering from the University of Colorado, an M.S. in geology from the University of New Mexico, and a Ph.D. from the California Institute of Technology. He is the W.M. Keck Foundation Professor for Resource Geology at the California Institute of Technology (CalTech), and his expertise is in petrology and geochemistry. Dr. Silver was a public works officer in the U.S. Naval Civil Engineer Corps from 1945 to 1946 and held several positions at the U.S. Geological Survey before he joined CalTech. He has served on numerous National Research Council committees, including his current membership of the Commission on Physical Sciences, Mathematics, and Applications. Dr. Silver is a member of the National Academy of Sciences.

CONSULTANTS

CHOPPIN, Gregory R.—Dr. Choppin received a B.S. in chemistry from Loyola University, New Orleans, and a Ph.D. from the University of Texas, Austin. He is currently the R.O. Lawton Distinguished Professor of Chemistry at Florida State University. His research interests involve the chemistry of the f-elements, the separation science of the f-elements, and concentrated electrolyte solutions. During a postdoctoral period at the Lawrence Radiation Laboratory, University of California, Berkeley, he participated in the discovery of mendelevium, element 101. His research and educational activities have been recognized by the American Chemical Society's Award in Nuclear Chemistry, the Southern Chemist Award of the American Chemical Society, the Manufacturing Chemist Award in Chemical Education, a Presidential Citation Award of the American Nuclear Society, and honorary D.Sc. degrees from Loyola University and the Chalmers University of Technology (Sweden).

DePAOLO, Donald J.—Dr. DePaolo earned a B.S. with honors from the State University of New York, Binghamton, and a Ph.D. from the California Institute of Technology. He is professor of geochemistry and director of the Center for Isotope Geochemistry at the University of California, Berkeley. Prior to arriving at Berkeley in 1988, Dr. DePaolo held a professorship at the University of California, Los Angeles. He is a recipient of the F.W. Clarke Medal of the Geochemical Society, the J.B. MacElwane Award of the Geophysical Union, and the Mineralogical Society of America Award. He is a member of the National Academy of Sciences.

HORNBERGER, George **M.**—Dr. Hornberger received an undergraduate degree in civil engineering but subsequently trained as a hydrologist at Stanford University, where he was awarded a Ph.D. in 1970. Dr. Hornberger is currently the Ernest H. Ern Professor of Environmental Sciences at the University of Virginia. He joined the University of Virginia's Environmental Sciences Department in 1970 and served as its chairman from 1979 to 1984. Dr. Hornberger has been the recipient of numerous awards, including election to the first group of fellows of the Association for Women in Science. He was cited for "exemplary commitment to the achievement of equity for women in science and technology." Dr. Hornberger received the John Wesley Powell Award from the U.S. Geological Survey and is also a member of the American Geophysical Union. He is the editor of *Water Resources Research*, the nation's premier journal for publications in the hydrological sciences. He was elected to the National Academy of Engineering in 1996.

APPENDIX D

SUPPLEMENTARY STATEMENT

Dr. Harold Lewis
January 10, 1997

I am uncomfortable about being forced to dissent from the committee's consensus report, but would be even more uncomfortable to accede to the temptation and pressure to sign a report that misses the mark. This is not to say anything negative about the committee chairman, for whom I have great respect—I simply seem to stand at one end of a spectrum of committee views on some important issues. The chairman was responsible for forging a consensus, and did so with patience and skill. I have often told students that to be in a minority doesn't make you wrong, but it does get you outvoted.

The committee was charged to say how basic research can help the Department of Energy (DOE), how basic research can add value to cleanup efforts, what kinds of technical challenges would benefit from basic research, what fields of research might be the most promising, and the like. It did none of this, concentrating its efforts in minute detail on micromanagement issues. The report recommends fellowships, scholarships, meetings, peer reviews, listings of publications, compilations of results, and the like—the cleanup problems require more than programmatic niceties. The Environmental Management Science Program (EMSP) is aimed at the real cleanup problems.

The program had as its origin the Appropriations Bill Conference Report of the 104th Congress, which expressed the hope that basic science research might help "to ultimately reduce cleanup costs." No other objective is mentioned, and the language makes clear that the concern is that current methods are too expensive, and are, by the way, also ineffective. This has been said by many, and is not new. It is even true. But the tasking to the committee from DOE did not list cost reduction as an objective, and the committee was left with the unenviable job of devising a set of objectives for the basic research program that is itself supposed to provide new ideas for a cleanup program pursuing its own unidentified objectives. The best the committee could do with this central question (which it was tasked to answer) was to recommend that the DOE develop a near-term science plan "from existing Department documents," and a long-term one by consulting with its "problem holders." Chapter 3 purports to describe how this can be done, but instead jumps into the Government Performance and Results Act (GPRA), and finally recommends that "the Department develop a science plan for the EMSP." Careful reading of Chapter 3 reveals that it never says how, or offers any but procedural direction. DOE needs help on substance, not procedures. Somehow, I would have expected more from an Academy committee.

This is not a trivial matter—it is central to the chance of success of EMSP. The logic that lies behind the congressional report, and appears elsewhere in many places, is that DOE has badly mismanaged this enormously expensive program, and that something has to be done to control the costs, now estimated in the hundreds of billions of dollars, over decades. The congressional love for basic research did not derive from any clear sense of how it could help, but from the foreboding (again shared by many) that the program is doomed unless something new is added. The only "something" available is basic research, which has the potential to generate useful new knowledge. The job of deciding how a basic research program could be structured to help was left to DOE (the very organization the Congress said was not paying enough attention to the subject), and DOE turned to the Academies, who are, in my view, letting them down. There is no substantive advice in this report to suggest how basic research can help, or how a program of needs-driven research can be kept basic yet applicable—there are lists of who should meet with whom, and how often.

Of course the problem goes far beyond the EMSP. Basic research in support of an objective can only be directed through awareness on the part of the investigators of what those objectives are, and an appropriate system of rewards. (Technology development is different—specifications can be set and enforced.) If anyone knows the ultimate objectives of the cleanup program, that wisdom has been kept marvelously secret. How then can a directed research program spring up spontaneously in the DOE community? Directed at what? And without a compass. In the cases I know in which basic research has led to technological advances of direct benefit to the sponsors of the research it has been because the investigators worked side by side with the potential users, and had the motivation to help. (The classic examples are the Bell Laboratories of old, the National Bureau of Standards (NBS) of old—now the National Institute of Standards and Technology (NIST), the National Institutes of Health (NIH), the better of the national laboratories, and so forth.) It will not be easy to direct basic research (by definition undirected) toward an objective, especially in the academic world, and some ideas from the Committee, whether or not original, would have been helpful. Lists of who should be consulted are not.

Finally, the congressional direction to DOE was specifically to have the program managed by the Office of Energy Research, but DOE opted instead for a two-headed structure composed of ER and EM, the latter precisely the organization responsible for the current unsuccessful program. The first program solicitation was managed by having ER review proposals for their scientific quality (using standards not revealed to the committee—we were told who won, but not who lost), and EM for "relevance," again using secret standards. Despite many requests, the committee was not given enough information to learn the criteria used to separate the winners from the losers in the first solicitation, but clearly each office had veto power, and EM the last word. I do not see how EM can be expected to suddenly be able to judge the relevance of a basic research proposal that deals with a truly novel approach to environmental management, when there has been no evidence of that skill in the past. And novelty is what the entire program is designed to produce—incremental improvements will not cut the mustard. ("Breakthrough" is the buzzword used in the report.) In truth, I believe that the committee's acceptance of this two-headed monster comes in large measure from the view that without the power and the associated sense of ownership, EM would drag its feet, and the program would die. If that is the case, it is no basis for condoning the shotgun wedding, and it is the Secretary's job to make the appropriate adjustments. (The committee recommendation here is for a single manager, reporting to the Under Secretary, but institutionally embedded in both offices. That is an improvement over the prior stance, to simply accept the monster into the family.) I believe that research should be managed as research (as the Congress intended originally). The research might be less closely tied to the current aspirations of EM that way, but will surely not be the finest basic research if it is even partially managed by people whose immediate objectives and career advancement considerations lead in other directions. Most specific basic research efforts do not pass a time-weighted cost-benefit test—it is only in the aggregate, over the long term, that basic research pays off in applications.

Let there be no mistake: I am a working scientist, and believe deeply in the power of basic research to provide the truth that sets us free. And it is even true that sometimes that truth has revolutionary applicability to the betterment of life (we remember those cases selectively, and with pleasure). Further, I agree that an expenditure of $50 million is trivially justifiable in this context. It is a gamble that is well worth taking—I have no difference at all with the committee on this point. But as now directed it is bound to suffer from the same disease that afflicts the cleanup program itself—lack of rationale and direction. It is a pipedream to believe that the finest scientists in the country will flock to the cleanup problem just because some money is available. (Besides, DOE and the committee have acted as if it were self-evident that they should. If that is obvious, I am obtuse. There are competing values.)

I think that the country, and perhaps even DOE, would have benefited from a deeper look at the rationale for EMSP, leading to a clearer view of how it should be organized and integrated into the DOE structure. Instead the committee chose an auditing approach that avoids the deep and fundamental questions, while micromanaging DOE on the others. As I read the charge to the committee, it was indeed asked some of the hard questions. It did not deliver.

APPENDIX E

RESPONSE TO SUPPLEMENTARY STATEMENT IN APPENDIX D

Dr. John Ahearne
February 14, 1997

Dr. Lewis correctly charges (Appendix D) that the Department of Energy's Environmental Management Program (DOE-EM) does not have a set of clear objectives. This is a point made forcefully by several previous National Research Council (NRC) reports,[1] on one of which I was a member. I agree that setting out such clear objectives would be of great value, not just for the EMSP, but for the overall EM program. However, this small study is not the place to take on this major task. Perhaps another NRC committee can be chartered and funded to do so— this is a major task, which must include examining whether changes will be needed to federal legislation (e.g., the Resource Conservation and Recovery Act; the Comprehensive Environmental Response, Compensation, and Liability Act; and the Federal Facilities Compliance Act), as well as negotiated agreements among states, the Environmental Protection Agency (EPA), and the DOE. The committee concluded that, even in the absence of such objectives, it is possible to fund basic science that may contribute significantly to meeting whatever objectives are finally agreed upon.

Dr. Lewis also disagrees with the committee's conclusion that the program should be a joint EM-ER program. The committee discussed this issue at length. While having some sympathy for Dr. Lewis's view that research is best left to the research community to administer, the

---

[1]National Research Council, 1995, Improving the Environment: An Evaluation of DOE's Environmental Management Program (Washington, D.C.: National Academy Press); National Research Council, 1996, Barriers to Science: Technical Management of the Department of Energy Environmental Remediation Program (Washington, D.C.: National Academy Press); National Research Council, 1996, Environmental Management Technology-Development Program at the Department of Energy: 1995 Review (Washington, D.C.: National Academy Press).

committee concluded that to ensure a working relationship between the researchers and those who own the problems, a joint program is better. The management solution we recommend is the committee's conclusion on how to best ensure that this relationship will work.

Therefore, much as I like and respect Dr. Lewis, I believe this report does provide DOE with substantial and significant advice on making the EMSP a viable program.

APPENDIX F

INITIAL ASSESSMENT REPORT

BUILDING AN EFFECTIVE ENVIRONMENTAL MANAGEMENT SCIENCE PROGRAM:

INITIAL ASSESSMENT

Committee on Building an Environmental Management Science Program
Virtual Commission on Environmental Management Science
National Research Council

NATIONAL ACADEMY PRESS
Washington, D.C. 1996

NOTICE: The project that is the subject of this report was approved by the Governing Board of the National Research Council, whose members are drawn from the councils of the National Academy of Sciences, the National Academy of Engineering, and the Institute of Medicine. The members of the committee responsible for the report were chosen for their special competencies and with regard for appropriate balance.

This report has been reviewed by a group other than the authors according to procedures approved by the Report Review Committee consisting of members of the National Academy of Sciences, the National Academy of Engineering, and the Institute of Medicine.

The work was sponsored by the U.S. Department of Energy, Contract No. DE–FC01–94EW54069/R. All opinions, findings, conclusions, and recommendations expressed herein are those of the authors and do not necessarily reflect the views of the Department of Energy.

Additional copies of this report are available from:

National Research Council
Virtual Commission on Environmental Management Science
2101 Constitution Avenue, N.W., HA 456
Washington, DC 20418
202-334-3066

Copyright 1996 by the National Academy of Sciences. All rights reserved.

Printed in the United States of America

COMMITTEE ON BUILDING AN ENVIRONMENTAL MANAGEMENT SCIENCE PROGRAM

JOHN F. AHEARNE, *Chair,* Sigma Xi, and Duke University, Research Triangle Park, North Carolina
EDWARD M. ARNETT, Duke University, Durham, North Carolina
STANLEY I. AUERBACH, SENES Oak Ridge, Inc., Oak Ridge, Tennessee
EDWARD J. BOUWER, Johns Hopkins University, Baltimore, Maryland
JOHN I. BRAUMAN, Stanford University, California
NAOMI H. HARLEY, New York University Medical Center, New York
DEREK R. LOVLEY, University of Massachusetts, Amherst
GENE G. MANNELLA, Gas Research Institute (retired), Potomac, Maryland
NORINE E. NOONAN, Florida Institute of Technology, Melbourne
LEON T. SILVER, California Institute of Technology, Pasadena

Consultants

GREGORY R. CHOPPIN, Florida State University, Tallahassee
DONALD J. DePAOLO, University of California, Berkeley
GEORGE M. HORNBERGER, University of Virginia, Charlottesville

Staff

KEVIN D. CROWLEY, *Study Director*[*]
TAMAE MAEDA WONG, *Senior Staff Officer*[†]
SUSAN B. MOCKLER, *Research Associate*[*]
PATRICIA A. JONES, *Project Assistant*[*]
ERIKA L. WILLIAMS, *Project Assistant*[*]
JOSHUA A. CHAMOT, *Intern*[*]

---

[*]Board on Radioactive Waste Management
[†]Board on Chemical Sciences and Technology

iii

VIRTUAL COMMISSION ON ENVIRONMENTAL MANAGEMENT SCIENCE

PERRY L. McCARTY, *Chair,* Stanford University, California
RICHARD A. CONWAY, Union Carbide Corporation, South Charleston, West Virginia
DONALD J. DePAOLO, University of California, Berkeley
DAVID J. GALAS, Darwin Molecular Corporation, Bothell, Washington
MICHAEL C. KAVANAUGH, ENVIRON Corporation, Emeryville, California
ROYCE W. MURRAY, University of North Carolina, Chapel Hill

Staff

STEPHEN RATTIEN, *Executive Director*

The Committee on Building an Environmental Management Science Program is a joint activity of the Commission on Engineering and Technical Systems; Commission on Geosciences, Environment, and Resources; Commission on Life Sciences, and Commission on Physical Sciences, Mathematics, and Applications.

The National Academy of Sciences is a private, nonprofit, self-perpetuating society of distinguished scholars engaged in scientific and engineering research, dedicated to the furtherance of science and technology and to their use for the general welfare. Upon the authority of the charter granted to it by the Congress in 1863, the Academy has a mandate that requires it to advise the federal government on scientific and technical matters. Dr. Bruce Alberts is president of the National Academy of Sciences.

The National Academy of Engineering was established in 1964, under the charter of the National Academy of Sciences, as a parallel organization of outstanding engineers. It is autonomous in its administration and in the selection of its members, sharing with the National Academy of Sciences the responsibility for advising the federal government. The National Academy of Engineering also sponsors engineering programs aimed at meeting national needs, encourages education and research, and recognizes the superior achievements of engineers. Dr. William A. Wulf is interim president of the National Academy of Engineering.

The Institute of Medicine was established in 1970 by the National Academy of Sciences to secure the services of eminent members of appropriate professions in the examination of policy matters pertaining to the health of the public. The Institute acts under the responsibility given to the National Academy of Sciences by its congressional charter to be an adviser to the federal government, and upon its own initiative, to identify issues of medical care, research, and education. Dr. Kenneth Shine is president of the Institute of Medicine.

The National Research Council was organized by the National Academy of Sciences in 1916 to associate the broad community of science and technology with the Academy's purposes of furthering knowledge and advising the federal government. Functioning in accordance with general policies determined by the Academy, the Council has become the principal operating agency of both the National Academy of Sciences and the National Academy of Engineering in providing services to the government, the public, and the scientific and engineering communities. The Council is administered jointly by both Academies and the Institute of Medicine. Dr. Bruce M. Alberts and Dr. William A. Wulf are chairman and interim vice-chairman, respectively, of the National Research Council.

PREFACE

This is the first of three reports by the Committee on Building an Environmental Management Science Program. The committee was established by the National Research Council to help the Department of Energy's Office of Environmental Management improve the effectiveness of its Environmental Management Science Program—a mission-directed, basic research program to support cleanup of the nation's nuclear weapons complex. The department announced this program in a Federal Register Notice in February 1996 and received more than 800 proposals from researchers at universities, national laboratories, and industry. The department is in the final stages of proposal review and expects to make award decisions in July 1996. In this initial assessment, the committee has restricted its findings and recommendations to the department's near-term needs as it completes the review of these proposals and develops the FY 1997 program plan. These near-term issues are well represented by the questions that constitute the statement of task for this first committee report:

- How can basic research be used to help DOE-EM to complete its mission successfully in the next few decades?
- How can a basic research program help add value to DOE-EM's cleanup efforts?
- What kinds of technical challenges would likely benefit from a program in basic research?
- How can the research program take advantage of the unique capabilities of U.S. universities and federal labs?
- How can the research program take advantage of research efforts and capabilities in other DOE programs and other federal agencies?
- What, if any, additional areas of research should be included in the FY 1997 program announcement as the DOE EMSP evolves?

The committee's future reports will address the longer-term science and management needs of this program and will be issued later this year.

CONTENTS

SUMMARY

In 1995, the 104th Congress directed the Department of Energy (DOE; see Appendix E for list of acronyms) to establish a basic research program to support its mission to clean up the nation's nuclear weapons complex. DOE established the Environmental Management Science Program (EMSP) in response to this mandate. This program is managed jointly by the department's Offices of Energy Research (ER) and Environmental Management (EM) and is designed to bridge the gap between "fundamental research" and "needs-driven applied research" in order to promote the development of new and improved cleanup technologies.

At the request of the DOE, the National Research Council established the Committee on Building an Environmental Management Science Program to advise DOE on ways to increase the effectiveness of this new research program. This report, the first of three that will be issued by the committee over the next seven months, provides an initial assessment of the EMSP that focuses on the fiscal year (FY) 1996 proposal competition and the FY 1997 program plan.

Given the size, scope, and long-term nature of the cleanup mission—DOE estimates that this effort will cost $230 billion and require 75 years—the committee views the establishment of this mission-directed, basic research program as both an urgent and a prudent investment for the nation. Although the EMSP will not solve all of EM's cleanup problems, a properly structured and managed program could help address many of EM's technical challenges by stimulating the development of new waste characterization, remediation, and management technologies or reducing the uncertainties in the application of current technologies; by enabling the development of new methods to reduce the volume or toxicity of secondary wastes; and by providing a better understanding of risk to help prioritize cleanup activities and reduce hazards to people and the environment.

The DOE faces at least three significant challenges in establishing a basic research program that has real long-term value to the cleanup mission:

1

(1) <u>Attracting the best researchers to the program:</u> Many of the nation's top scientists and their graduate students currently are not involved in research of direct relevance to the EMSP, although they have the background and skills necessary to do work at the forefront in this area. Fundamentally, the DOE will need to demonstrate a long-term commitment to this research program before scientists will redirect their research and graduate student training activities to the program's concerns.

(2) <u>Obtaining the best research ideas:</u> In order to obtain the "best" (i.e., meritorious and relevant) basic research in the EMSP, researchers must become knowledgeable of EM's research needs, both its generic needs and its site-specific needs. Additionally, a process must be established for identifying meritorious proposals for funding and, as a corollary, a process for providing useful feedback to researchers who are unsuccessful in obtaining funding for their research ideas.

(3) <u>Transferring research results to potential research users:</u> For the EMSP to contribute to the long-term cleanup mission, effective mechanisms must be found to transfer the results of the research to the "users"—technologists in government, industry, and academia who can utilize this knowledge to develop new or improved cleanup methods.

The DOE initiated the EMSP on an accelerated schedule in response to congressional actions, and the 1996 proposal competition is well under way. The review process that DOE has outlined to the committee seems reasonable and should lead to the support of scientifically meritorious proposals that are relevant to the long-term cleanup mission. The committee offers the following advice to DOE as it completes the review process:

• In making award decisions in this first round, DOE should focus first on scientific merit and then on potential relevance to the cleanup mission, and should place less emphasis on the "anticipated" institutional funding allocations announced in the program notice. In this regard, DOE should relax its initial allocation of $20 million for proposals from national laboratories and $20 million for proposals from academia and industry to the extent allowed by the law, and, instead, should allocate funds to support the most scientifically meritorious and relevant work, regardless of the institution of origin. Similarly, in evaluating the merit of collaborative research proposals, DOE should focus on the potential value added by the nature and scope of the proposed collaborations, not only on the number or size of institutional or researcher commitments to a particular project.

• Successful proposals should be funded fully "up front" to help ensure the stability and continuity of the research projects and to establish a solid foundation on which a stable, long-term program can be built.

The committee believes that the FY 1997 program plan will be a major—and perhaps the defining—step in shaping this program. In particular, it will be important for DOE to establish a focus for the EMSP that builds on, but does not duplicate or divert funding from, existing ER programs in order to improve the usefulness of the research to the long-term cleanup mission. To ensure the program's long-term success, the committee recommends that DOE

• with the advice of the research and research-user communities, prepare concise written technical summaries of the critical barriers to the solution of cleanup problems and basic research needs for wide circulation to the research community;
• postpone until later this year the release of the 1997 program notice until it has had time to identify and incorporate the "lessons learned" from the FY 1996 proposal competition and to think more carefully, using the advice of this committee where appropriate, about how the program should be structured and managed; and
• seek to increase the budget for this program to FY 1996 levels, recognizing that the additional funds are likely to be reallocated from existing programs within DOE-EM, in order to provide level funding, which is necessary to establish a stable, long-term research program.

In the committee's judgment, the long-term success of this program is highly dependent on the continuing partnership between EM, which understands the cleanup problems and research needs, and ER, which, through its mission to manage the department's basic research programs, understands how to select and manage research. The committee strongly endorses the efforts made by EM and ER staff to work together and encourages them to continue their efforts to build an effective Environmental Management Science Program.

INTRODUCTION AND BACKGROUND

The Department of Energy's (DOEs) Environmental Management Science Program (EMSP) was created by mandate of the 104th Congress[1] to focus the nation's research infrastructure on the department's environmental cleanup mission:

> The conferees agree with the concern expressed by the Senate that the Department [of Energy] is not providing sufficient attention and resources to longer term basic science research which needs to be done to ultimately reduce cleanup costs. The current technology development program continues to favor near-term applied research efforts while failing to utilize the existing basic research infrastructure within the Department and the Office of Energy Research. As a result of this, the conferees direct that at least $50,000,000 of the technology development funding provided to the environmental management program in fiscal year 1996 be managed by the Office of Energy Research and used to develop a program that takes advantage of laboratory and university expertise. This funding is to be used to stimulate the required basic research, development and demonstration efforts to seek new and innovative cleanup methods to replace current conventional approaches which are often costly and ineffective.

A working partnership between the Office of Environmental Management (EM) and the Office of Energy Research (ER) was begun in 1994 to establish a basic research program focused on EM needs. The importance of basic scientific research to the cleanup mission has been established in several reports, most recently the report of the Galvin commission, entitled *Alternative Futures for the Department of Energy National*

---

[1] Public Law 104-46, 1995. The text is from the conference report that accompanied H.R. 1905 (Energy and Water Development Appropriation Bill).

Laboratories (DOE, 1995a), and the National Research Council (NRC, 1996) report entitled *Improving the Environment: An Evaluation of DOE's Environmental Management Program:*

> Probably the most important reason behind the slow pace of assessment and cleanup is the low quality of science and technology that is being applied in the field. . . . There is a lack of realization that many—and some experts believe most—existing remediation approaches are doomed to technical failure. Others would require unacceptable expenditures and much extended time to reach their stated objectives. . . . There is a particular need for long-term, basic research in disciplines related to environmental cleanup. . . . Adopting a science-based approach that includes supporting development of technologies and expertise . . . could lead both to reduced cleanup costs and smaller environmental impacts at existing sites and to the development of a scientific foundation for advances in environmental technologies. (DOE, 1995a, pp. 30, 40-41)

> EM has recently begun an effort to coordinate its technology-development efforts with the Office of Energy Research, which houses much of the Department's basic research and is the principal office for interaction with non-defense Department National Laboratories. . . . This type of linkage, including the defense-related laboratories, where much of the expertise in nuclear materials resides, is precisely what is called for . . . . The Department should extend this attempt to create partnerships to include the basic-research efforts in universities and industrial concerns that are developing technology or undertaking their own research. (NRC, 1996, p. 117)

The EMSP is a long-term research program designed to bridge the gap between fundamental research and needs-driven applied technology development (see Appendix A). The objective of this program is to generate new knowledge that will lead to less costly, more innovative cleanup technologies and will reduce risks to workers, the public, and the environment. An important focus of the program is the development of new knowledge to deal with problems that are intractable by using current technologies and to inspire "breakthroughs" in areas critical to the EM cleanup mission.

The first EMSP proposal announcement targeted to university and industry researchers was published in the February 9, 1996, Federal Register (Volume 61, No. 281; see Appendix B). As a result of this announcement, and a similar solicitation directed at national laboratory researchers, the program received about 2,200 preproposals and, subsequently, 810 full proposals on topics ranging from bioremediation to sensor development. DOE is now in the process of reviewing these proposals and expects to make awards later this year.[2] A description of the FY 1996 EMSP and review process is given in Appendix A.

In a letter to Bruce Alberts, President of the National Academy of Sciences (NAS), Under Secretary of Energy Thomas P. Grumbly requested the assistance of the NAS in advising DOE on ways to increase the effectiveness of this research program. The Committee on Building an Environmental Management Science Program was established under the auspices of the National Research Council (NRC) to undertake this work. During this 10-month study, the committee will issue three reports that address both the science and the management needs of the program.

The issues facing DOE in establishing and managing an effective EMSP are well represented by the questions that constitute the statement of task for this first committee report:

- How can basic research be used to help DOE-EM complete its mission successfully in the next few decades?
- How can a basic research program help add value to DOE-EM's cleanup efforts?
- What kinds of technical challenges would be likely to benefit from a program in basic research?
- How can the research program take advantage of the unique capabilities of U.S. universities and federal labs?
- How can the research program take advantage of research efforts and capabilities in other DOE programs and other federal agencies?
- What, if any, additional areas of research should be included in the FY 1997 program announcement as the DOE EMSP evolves?

---

[2]Of the $50 million allocated to this program in FY 1996, $20 million has been set aside to fund proposals from universities and industry, $20 million has been set aside to fund proposals from national laboratories, and $10 million has been set aside for administration and special project costs.

In addressing these questions in this first report, the committee has restricted its findings and conclusions to near-term needs of the EMSP, in order to provide timely advice to DOE for use in completing the review of this year's proposals and in developing the FY 1997 program, consistent with the committee's compressed schedule for information gathering and deliberation. Longer-term science and management needs of the program will be addressed in the second and third reports, which will be issued later this year. The project schedule is described later in this report.

Information used to develop this report was obtained by the committee during two meetings at which it received briefings from DOE, from university, national laboratory, and industry researchers (Appendix C) and from the committee's review of previous NRC and DOE reports relevant to this program.

THE DOE CLEANUP MISSION

Fifty years of nuclear technology and weapons development have produced both positive and negative legacies for the nation. Nuclear technology contributed to national security during the Cold War, but the treatment and disposition of radioactive and chemical wastes were a secondary concern to the production of nuclear weapons. These weapons production efforts have left the nation with contaminated soil, surface water, and ground water, as well as large volumes of radioactive and chemical wastes, that are a hazard to human health and the environment.

The DOE is the agency responsible for managing the nuclear weapons complex, including more than 120 million square feet of buildings and facilities and 2.3 million acres of land that were used for the research, production, and testing of nuclear weapons (DOE, 1995c). The department's cleanup challenge is huge in scope and includes[3] 3,700 contaminated sites in 34 states and territories; more than 100 million gallons of radioactive and mixed wastes stored in 322 tanks; 3 million cubic meters of radioactive or hazardous buried wastes; 250 million cubic meters of contaminated soils from landfills and plumes; more than 600 billion gallons of contaminated ground water; and about 1,200 facilities that require decontamination and decommissioning. As an example, there are approximately 215 million curies of radioactivity in the 177 storage tanks at the Hanford site (Gephart and Lundgren, 1995). Innovative characterization and remediation technologies will be required to characterize and stabilize this waste

---

[3]From written material received from DOE-EM at the first committee meeting.

over the long term to keep it from further contaminating the local environment.

Cleanup of the weapons complex is necessary to protect human and environmental health, but such cleanup will be difficult and expensive. Based on the use of existing technologies and cleanup approaches, DOE's current estimate of cleanup costs is $200 billion to $350 billion, with a midrange estimate of $230 billion, over 75 years (DOE, 1995b).[4] Of this total, DOE estimates that $112 billion will be spent for waste management, $65 billion for environmental restoration, $22 billion for nuclear material and facility stabilization, $12 billion for technology development, and the remainder for activities such as program management and planning and annual monitoring (DOE, 1995b). This estimate does not include costs for problems that DOE believes cannot be solved with current technologies, such as cleanup of the large volumes of contaminated soil and ground water that exist at many sites.

According to DOE, the most urgent and high-risk tasks are the stabilization and maintenance of a large number of nuclear facilities and materials (DOE, 1995b), including the prevention of material leaks, explosions, theft, terrorist attack, and avoidable radiation exposures. The inherent difficulties associated with the handling and storage of radioactive materials, in addition to the vast quantity and varied forms of this waste, suggest that comprehensive cleanup will be a formidable goal.

The DOE established the Office of Environmental Management (EM) in 1989 to manage this cleanup effort. Within this office, programs were established in environmental restoration, waste management, nuclear material and facility stabilization, and technology development and were charged with the following six goals (DOE, 1995b): (1) eliminate and manage urgent risks; (2) emphasize health and safety for workers and the public; (3) establish a system that is managerially and financially in control; (4) demonstrate tangible results; (5) focus technology development on identifying and overcoming obstacles to progress; and (6) establish a stronger partnership between DOE and its stakeholders (i.e., those groups that have a "stake" in the process and outcome of cleanup, including workers, regulators, and communities around the sites).

Many of EM's cleanup problems cannot be solved or even managed efficiently and safely with current technologies, in part owing to their

---

[4]As noted in *The 1995 Baseline Environmental Management Report* (DOE, 1995b), these estimates involve many uncertainties, and future estimates may change as more information becomes available. There are no independent estimates of the magnitude of cleanup costs.

tremendous size and scope. However, cleanup would benefit greatly from the involvement of basic researchers, as noted in a recent NRC report (NRC, 1996, pp. 6-7):

> In some circumstances, technologies and processes for safe and efficient remediation or waste minimization do not exist. In other cases, the development of new technology and processes might substantially reduce the costs of, or risks associated with, remediation and waste management. . . . In some cases, fundamental science questions will have to be addressed before a technology or process can be engineered. . . . There is a need to involve more basic science researchers in the challenges of the Department's remediation effort.

THE VALUE OF RESEARCH TO THE CLEANUP MISSION

The DOE-EM cleanup mission has been called the world's largest civil works project (e.g., Blush and Heitman, 1995; Zorpette, 1996) and is in many ways more demanding scientifically and technically than the effort to develop nuclear weapons, which began with the Manhattan Project. As noted in the previous section, the nation lacks the scientific and technical know-how to address many of the most pressing cleanup problems and is confronted with the prospect of spending large sums of taxpayer funds simply to prevent the further spread of contamination. A research program could add significant value to EM's cleanup mission by producing new knowledge that will stimulate the development of technologies and methods to improve the effectiveness and lower the costs and risks of cleanup.

As noted in the introduction of this report, Congress directed DOE to develop a science program that would utilize the "existing basic research infrastructure within the Department and the Office of Energy Research" and would take "advantage of [federal] laboratory and university expertise." EM already supports activities that could be classified as research or research and development (R&D) through its Office of Science and Technology (EM-50). The conference report language suggests that this new research program should support a kind of research that is distinctly different from that currently supported by EM-50.

The program notice (Appendix B) states that the objective of the program is to "'[b]ridge the gap' between broad fundamental research that has wide-ranging applicability . . . and needs-driven applied technology development. . . ." This program would probably be recognized by most scientists as a mission-directed, basic research program. The program is "mission-directed" in that research will be supported only in certain high-priority areas dictated by DOE's cleanup challenges. The program is "basic" in that it is focused on the investigation of fundamental physical, chemical, geological, and biological processes and phenomena, with no specific technology in mind and no established time horizon for payoff.[5]

The committee believes that a properly structured and managed mission-directed, basic research program can produce knowledge that would add significant value to EM's technology development efforts. Such knowledge, if properly applied, could help address the following technical challenges:

• Characterization, remediation, and management of radioactive and chemical wastes: Basic research may help stimulate the development of new technologies and reduce the uncertainties involved in the application of current technologies.

• Secondary wastes: Basic research may lead to the development of new methods to reduce the volume and toxicity of the secondary wastes generated by cleanup.

• Risk: Basic research may provide a better understanding of risk, which would help EM prioritize its cleanup activities and reduce hazards to workers, the public, and the environment.

The committee can imagine several specific cleanup problems that could be addressed through a focused program of basic research. Basic research in chemistry, for example, could stimulate the development of new instruments and analytical methods to help characterize the 55 million gallons of hazardous and radioactive wastes that exist in the tanks at the Hanford site. Similarly, basic research in geoscience and engineering science on flow and transport phenomena could lead to a better understanding of subsurface flow processes, which in turn could improve the effectiveness of

---

[5]Terms such as basic research are used frequently but seldom understood precisely. Good definitions of this and related terms are provided in *Allocating Federal Funds for Science and Technology* (NRC, 1995, p. 6), where basic research is characterized as research that "creates new knowledge; is generic, non-appropriable, and openly available; is often done with no specific application in mind; requires a long-term commitment."

efforts to stabilize and remediate contaminated ground water. Basic research in the biological sciences could stimulate the development of new or improved biological agents to break down chemical waste or sequester radioactive waste, thereby improving the effectiveness of waste treatment and reducing the volume of secondary wastes generated during cleanup.

The committee believes that the Department of Energy and the nation should view funding for the EMSP as a long-term investment that will provide payoffs over the life of the cleanup mission in terms of both lower risks (to workers, the public, and the environment) and costs and of improved effectiveness. This investment is not unlike the R&D investments made by successful for-profit, private-sector firms, which recognize that R&D is essential to long-term survival and prosperity. The committee notes that DOE's first-year investment in the EMSP is modest compared to many private-sector R&D efforts—the department's investment represents about 0.8 percent of EM's annual budget, and the total EM investment in R&D represents about 6.6 percent of its budget.[6] By comparison, "high-technology" manufacturing firms (e.g., computing, electronic, communication, instrumentation, and pharmaceutical firms) spend between about 7 and 12 percent of net sales on R&D.[7]

The committee emphasizes that DOE's investment in the EMSP will not solve all of its cleanup problems and needs to be viewed as "risky" in financial terms, in that there is no absolute guarantee of any quantifiable return and, moreover, it may be difficult to track precisely the returns on dollars invested. However, in the context of a long-term mission of EM, where many of the most serious remediation problems are technically challenging—and exorbitantly expensive to solve with current knowledge and technologies—the investment in basic research is viewed by the committee as both prudent and urgent. The risks inherent in supporting basic research in the EMSP are small in comparison to the potential payoffs.

---

[6]The total EM budget in FY 1996 was $6.1 billion. Of this total, $349.9 million was allocated to EM-50 to support technology development, and $50 million was allocated for the EMSP.

[7]Data on the R&D expenditures of manufacturing firms are for the year 1993 and are taken from *Science and Engineering Indicators—1996* (National Science Board, 1996). Data for R&D spending by the federal government and the private sector are not directly comparable because they are computed on different bases; nevertheless, they do allow for a rough comparison of relative efforts.

UTILIZING THE CAPABILITIES OF THE RESEARCH INFRASTRUCTURE

The EMSP is being established at a time of tumultuous change in the partnership between the scientific research enterprise and society. New public funds for scientific research are becoming scarce, and scientists are increasingly being held accountable for the benefits that their research conveys to society (NRC, 1993; Office of Science and Technology Policy, 1994). At the same time, the intellectual challenge of research on environmental problems, and the importance of such research to the nation, increasingly are being recognized by the nation's best scientists. A properly focused and managed EM scientific program could attract the nation's top researchers, promote the training of the next generation of environmental scientists, and thereby serve as an important driver for environmental research in the United States.

The strength of the U.S. research community lies in the depth and diversity of its talent and its institutions; this is particularly true in the disciplines relevant to DOE's cleanup mission. DOE, however, faces at least three significant challenges in bringing this considerable talent to bear in the EMSP and obtaining research that has long-term value to its cleanup mission:

1. attracting the best researchers,
2. obtaining the best research, and
3. transferring research results to potential research users.

The committee plans to devote considerable attention to these issues during the course of its study, and it offers some preliminary comments on these points in the following sections.

Attracting the Best Researchers

The objective of the EMSP is to foster "knowledge breakthroughs"[8] that will be of long-term value to cleanup of the weapons complex. Although a properly managed basic research program can produce such breakthroughs, it is difficult, if not impossible, to predict where these

---

[8]The committee uses the term "breakthrough" advisedly, because most advances in knowledge are incremental in nature.

will occur, and the breakthroughs themselves may not even be recognized until long after the research is completed. The committee believes that the EMSP is most likely to stimulate knowledge breakthroughs of value to DOE through a "bottom-up" process in which the nation's best scientists are encouraged to submit research proposals. Thus, the committee notes, and endorses, DOE's decision to encourage submission of proposals from researchers in a wide range of disciplines and institutions (Appendix B) in the FY 1996 program.

Many of the nation's top scientists and their graduate students currently are not involved in research of direct relevance to the EMSP, although they have the background and skills necessary to do work at the forefront in this area. Fundamentally, the DOE will need to demonstrate a long-term commitment to this research program before scientists will redirect their research and graduate student training activities to the program's concerns. The redirection of a research program is a significant undertaking with long-term career implications. It can require several years of sustained effort for one to become familiar with a new research field and conversant in its literature. In some cases, it can also require substantial financial commitments, both on the part of the scientists and their institutions, to upgrade equipment and facilities. The nation's top scientists will be unwilling to make such shifts without a high-level of confidence that funding will be available over the long term to support research and graduate student training.

The nation's best scientists can be found in a broad spectrum of research institutions—universities, industry, national laboratories, and other federal agencies—and these researchers and their institutions have unique strengths that can be tapped for the EMSP:

- National laboratory researchers: Many national laboratory researchers are familiar with the weapons complex and the cleanup mission, and they possess specialized knowledge, equipment, and analytical and monitoring capabilities. Many of these researchers also are experienced in working in large teams that may be useful to address certain types of multidisciplinary problems.

- Industry researchers: Industry researchers share many of the talents of their national laboratory counterparts—access to specialized knowledge and equipment, and experience in working in multidisciplinary team environments. Some also have a familiarity with the cleanup mission and problems. In addition, many industrial researchers have experience working on mission-directed research and working at the interface between research and application.

• <u>University researchers:</u> University researchers are at the forefront in many of the fundamental scientific disciplines—biology, chemistry, engineering, geoscience, and physics—where advances in knowledge are likely to provide large future payoffs to the cleanup mission. Through their training of graduate students, university scientists will produce the nation's future generations of researchers, which, if properly nurtured, could become a "committed cadre" of researchers for the EMSP.

• <u>Researchers at other federal agencies:</u> Many federal "mission" agencies have considerable research talent and capabilities in specific areas that are relevant to EM's research needs. Researchers at the U.S. Geological Survey (USGS), for example, are performing "cutting-edge" research on many problems related to ground water monitoring and remediation, and Environmental Protection Agency (EPA) researchers are at the forefront in certain areas of health effects research.

In addition, other nations are dealing with radioactive waste and chemical cleanup problems, and the international research community has expertise in both generic basic research and site-specific, problem-oriented research of potential value to the EMSP.

The long-term success and effectiveness of the EMSP will depend to a large extent on the degree to which the program is able to tap into this community of researchers, and a particular challenge for DOE will be to find ways to involve this community as the program evolves. In the near term, this community can be tapped by encouraging collaborative "networking" among researchers, which may or may not involve direct research funding from the program but could involve carefully targeted opportunities such as workshops, seminars, and fellowships. The committee notes that precedents for such collaborative activities already exist in many of DOE's programs. For instance, there is a long history of collaborations of university faculty and graduate students with national laboratory science groups. These collaborations were begun soon after the formation of the Atomic Energy Commission, a precursor agency to DOE, for the very reason that it was deemed essential to train and educate new researchers in the fields of science opened by atomic energy. Graduate and postgraduate training in collaboration with university faculty is a long-standing tradition at many DOE research laboratories. National laboratory researchers have also established productive working relationships with a variety of federal agencies.

The FY 1996 program notice (Appendix B) encourages collaborations among researchers in universities, national laboratories, and industry, where appropriate. The committee recognizes, and endorses in principle, the importance of collaboration between researchers, but points out that collaborations can extend beyond the university-industry-laboratory triad and can take a variety of forms—ranging from informal communication among researchers working on single-investigator projects, to teams of researchers working in close coordination on complex, multidisciplinary projects. The committee notes that much of the nation's best science continues to be done by single investigators working on individual projects. In order to build an effective EMSP, DOE must find ways to identify and encourage the appropriate types of value-added collaborations that will help it address the full range of its research needs. In future reports, the committee will consider ways to optimize the usefulness of collaborative activities to the EMSP.

Obtaining the Best Research

In order to obtain the "best" (i.e., meritorious and relevant) basic research in the EMSP, researchers must become knowledgeable of EM's research needs, both its generic needs and its site-specific needs. The FY 1996 program notice (Appendix B) lists a broad range of generic research needs and serves as a good starting point for informing the research community. Some of ER's reports and research solicitations—for example, *Basic Research for Environmental Restoration* (DOE, 1990) and the program solicitation *Natural and Accelerated In-Situ Bioremediation Program* (DOE, 1995d)—can also serve this function. Additionally, DOE has developed a great deal of written documentation on cleanup needs that could also serve to inform the research community—for example, *Estimating the Cold War Mortgage: The 1995 Baseline Environmental Management Report* (DOE, 1995b); the focus area reports (DOE, 1995e-i); and more problem-specific reports such as the *Hanford Tank Cleanup: A Guide to Understanding the Technical Issues* (Gephart and Lundgren, 1995). Much of the information in these reports, however, addresses near-term needs and is not organized or written to be easily accessible to researchers.

To improve the communication of EM's problems to researchers, the committee recommends that DOE prepare concise written technical summaries of its basic research needs for the research community. Such summaries should contain information on the critical barriers to the solution of EM's problems, arranged both by site and by problem focus. In preparing these summaries, the DOE should seek the advice of the research and research-user communities to ensure that the summaries reflect EM's highest-priority needs and that the research questions are framed properly. These summaries should be produced for wide circulation to the research community and should be updated as appropriate to reflect current needs.

The committee also recommends that DOE consider other ways to give researchers information about contaminated sites, for example, by providing site-specific briefings to researchers on problems and needs so that they can familiarize themselves with the cleanup challenges and establish lines of communication with the "problem holders" and potential users of their research, or by supporting informal interactions between researchers at national laboratories and those in universities who are studying similar problems, through mechanisms such as workshop and seminar programs at cleanup sites or national laboratories.

In soliciting research proposals for the EMSP, DOE should take advantage of the potential value added from field research conducted at non-DOE sites. A number of DOE's waste problems are "generic" in nature, such as ground water contamination by chlorinated solvents, petroleum hydrocarbon mixtures, and certain heavy metals. Opportunities for field-scale research on these problems exist at sites managed by the USGS, EPA, and the Department of Defense (DOD), among others. Research projects that utilize appropriate non-DOE "testbeds" can provide understanding that can be transferred directly to cleanup of the weapons complex.

Another significant management challenge for getting the best research is establishing a process for identifying meritorious proposals for funding and, as a corollary, a process for providing useful feedback to researchers who are unsuccessful in obtaining funding for their research ideas. DOE faces a dual challenge in this effort: it must have a process that can identify research ideas that are both *scientifically meritorious* and *relevant* to EM's cleanup mission. Peer review,[9] of course, should be an integral part of identifying scientifically meritorious proposals, and the committee notes that this process is being used by DOE to evaluate the proposals it received in FY 1996 (Appendix A). The best process for establishing relevance to cleanup is less clear to the committee. The committee comments on this process in more detail later in this report.

---

[9]The committee defines peer review as review by scientists who work in the same or related research fields and who are not employed by the funding agency. Such peer review is used in many of ER's programs and at other agencies such as the National Institutes of Health and the National Science Foundation. See NRC (1995, p. 25) for additional discussion of the peer review process.

Transferring Research Results to Potential Research Users

For the EMSP to contribute to the long-term cleanup mission, effective mechanisms must be found to transfer the results of the research to the "users"—technologists in government, industry, and academia who can utilize this knowledge to develop new or improved cleanup methods. An important component of this transfer process is the open publication of research results using the traditional venues of national and international scientific meetings and peer-reviewed journals. These conventional publication outlets work well for communication of research results within the scientific community, but they may work less well for reaching those involved in technology development. In its future reports, the committee will consider the potential benefits of more dedicated dissemination activities—for example, workshops that bring together researchers and the users of research, and special DOE or independent publications to announce research results that can be developed and implemented rapidly to give valuable near-term technology payoffs. The committee will pay close attention to the balance between the costs and benefits of these special dissemination activities, given the budget and human resource limitations for the EMSP.

COORDINATION WITH OTHER FEDERAL AND NONFEDERAL RESEARCH PROGRAMS

The committee's statement of task directed it to address the question of how the EMSP could take advantage of research efforts and capabilities in other DOE programs and other federal agencies. The committee offers some preliminary comments directed to this issue in this section.

The EMSP was created very quickly by DOE in response to congressional mandate, and it is the committee's impression that the program was established without much planning for coordination with existing ER programs—such as the "core" research programs in basic energy sciences or cross-cutting research programs such as the Natural and Accelerated In-Situ Bioremediation (NABIR) program (DOE, 1995d). These ER programs are vital to the department's long-term mission and are an important part of the nation's basic research portfolio. The committee believes that it will be important for DOE to establish a focus for the EMSP that builds on—but does not duplicate or divert funding from— these existing ER programs in order to improve the usefulness of the research to the long-term cleanup mission.

The DOE must also become cognizant of other federal and nonfederal research efforts in order to obtain access to a broader researcher and knowledge base, to improve the focus of the EMSP, and to reduce needless duplication. The committee is aware of several research programs that are potentially relevant to the EMSP, including the following examples:

- The joint DOE, EPA, National Science Foundation (NSF), and Office of Naval Research program in bioremediation.
- The joint EPA and NSF program in water and watersheds.
- EPA research programs addressing risk, ecological assessment, and hazardous waste.
- NSF "core" research programs in the physical and social sciences, and NSF interdisciplinary programs focused on environmental problems.
- Research programs of the National Institute of Environmental Health Sciences (part of the National Institutes of Health complex).
- DOD research programs.
- Research sponsored by nonfederal organizations (e.g., the Gas Research Institute).

The committee will be gathering information on such programs and will comment on effective coordination strategies in future reports.

FY 1996 PROGRAM PRIORITIES AND SOLICITATION

The process for reviewing proposals and making awards in the FY 1996 EMSP is well under way. Congressional action required DOE to initiate the FY 1996 program on an accelerated schedule, which may not have allowed researchers adequate time to educate themselves about EM's cleanup problems and research needs or to prepare proposals that were fully responsive to, or addressed the full breadth of, problem areas outlined in the program notice (Appendix B). The FY 1996 schedule also presented significant challenges to both ER and EM in managing the review process (Appendix A). Future competitions (in FY 1997 and beyond) offer important opportunities to reflect on the experience of the FY 1996 program and to give further careful consideration to both the content and the process of the EMSP.

In the FY 1996 program notice, DOE provided several criteria for evaluating proposals and making awards (Appendix B), including (1) scientific and technical merit (e.g., assessment of the potential for addressing problems identified in the program notice and of relevance to the cleanup mission) and (2) appropriateness of the approach. In making award decisions in this first round, the committee recommends that DOE focus first on scientific merit and then on potential relevance to the cleanup mission and place less emphasis on the "anticipated" institutional funding allocations announced in the program notice (Appendix B; see also footnote 2). In this regard, the committee knows of no scientific justification for DOE's allocation of $20 million for proposals from national laboratories and $20 million for proposals from academia and industry—and in fact believes that this allocation could prevent DOE from funding the most meritorious and relevant proposals. The committee strongly recommends that the DOE relax this allocation to the extent allowed by the law, and award funds to support the most scientifically meritorious and relevant work, regardless of the institution of origin. Additionally, when evaluating the merit of collaborative research proposals, the committee encourages the DOE to focus on the potential value added by the nature and scope of the proposed collaborations, not only on the number or size of institutional or researcher commitments to a particular project.

The review process that DOE outlined for the FY 1996 program (Appendix A) seems reasonable to the committee, particularly given the short time frame for decision making. The original plan called for external reviews to assess scientific and technical merit by using panels of scientists. Following external review, EM program managers were to review the proposals for relevance and to prioritize them for EM management.[10] The committee had some initial concerns that this process could have diluted the quality of the science because the "relevance" review appeared to be a somewhat separate process. Based on a briefing it received at its May meeting, however, the committee now understands that the proposals are to be judged first for scientific quality; the group of most meritorious proposals then will be reviewed for relevance by knowledgeable EM managers assisted by ER staff (Appendix A). The committee endorses such a joint effort because it will serve to keep scientific and technical merit "front and center" in the review process while giving proper weight to the important criterion of relevance. The committee intends to revisit the review process

---

[10]As noted in a memorandum dated May 6, 1996, from C.W. Frank to Deputy Assistant Secretaries and Assistant Managers for Environmental Management.

in its entirety after completion of the FY 1996 competition and to suggest improvements if appropriate.

The committee recommends that, once award decisions are made, successful proposals be funded fully "up front." The committee recognizes that full funding may, indeed, decrease the absolute number of proposals that can be supported in this round but nevertheless believes that full funding is necessary for the following four reasons:

(1) To establish a solid foundation on which a stable, long-term program can be built.

(2) To ensure that projects funded in the first round will be completed on schedule and that research results will be available to potential users in the near term.

(3) To free-up funding for new starts in FY 1997, which, as noted previously, will be *essential* to convince the nation's best scientists to redirect their current research efforts in order to become familiar with EM's research needs and to submit research proposals.

(4) To provide opportunities to support high-quality proposals in the FY 1997 program. The committee expects that the proposals submitted to the program in FY 1997 will be of higher quality, on average, than proposals in this year's competition, because researchers will have more time to learn about EM's needs and prepare proposals.

In short, full funding will accelerate the establishment of what the committee has referred to as a "committed cadre" of the nation's top researchers—scientists knowledgeable of EM's problems and needs who produce research results that have long-term value to the cleanup mission.

The committee understands that there may be special administrative issues with regard to providing full funding for proposals where the principal performer is a national laboratory. The committee believes, however, that mechanisms can and must be found to enable full funding for all performers.

The committee also believes that it will be important for DOE to review the progress of the projects it funds on a periodic basis to ensure that they remain focused and that appropriate progress is being made. The committee notes that some ER programs have established processes for such reviews and a process for discontinuing support of unproductive projects. The committee will examine the usefulness of these and other review mechanisms in future reports.

For researchers whose proposals are unsuccessful in attracting funding in this first competition, DOE should make a special effort to encourage their continued participation by providing written feedback (e.g., written reviews and summaries of panelist comments) that will help improve their future submissions to the program. The EMSP activity is fragile, and DOE needs to take special care not to discourage well-qualified and competent researchers in this formative stage. DOE should continue and even expand its outreach efforts to improve the understanding and appreciation of the magnitude of EM problems and EM research priorities by the national and even international scientific and technical communities. As noted earlier, the committee will address outreach activities in a future report.

FY 1997 PROGRAM

The EMSP has been jointly implemented by the DOE Offices of Environmental Management and Energy Research, but it is not yet clear to the committee what the long-term management structure of the program will be. The committee views this partnership between EM and ER as being vital to the long-term success of the EMSP, because it combines ER's expertise in research selection and management with EM's knowledge of cleanup problems and research needs. In the committee's view, the program should continue to build on the strengths of these two DOE offices to identify meritorious long-term research that is relevant to the EM cleanup mission.

The FY 1997 program plan will be a major—and perhaps the defining—step in shaping the scope and ensuring the success of the EMSP. Consequently, the committee strongly recommends that DOE postpone, until later this year, the release of the 1997 proposal solicitation[11] until it has had time to identify and incorporate the "lessons learned" from the FY 1996 proposal competition and to think more carefully, using the advice of this committee where appropriate, about how the program should be structured and managed.

As it develops FY 1997 program plans, DOE also needs to think carefully about funding levels. The committee believes that level funding is a minimum requirement to establish a stable, long-term research program that attracts highly relevant proposals from the nation's top researchers

---

[11]The committee learned at its May meeting that DOE intends to issue the FY 1997 program notice in September.

and notes with concern that the FY 1997 funding request for this program is only $38 million,[12] $12 million less than is available in FY 1996 (see footnote 2). The committee strongly recommends that DOE seek to increase the budget for this program to FY 1996 levels, recognizing that the additional funds are likely to be redirected from existing programs within DOE-EM.

In its future meetings and reports, the committee will address several issues of direct relevance to development of the FY 1997 EMSP, including the following:

- Articulation of research needs: The committee's statement of task directed it to advise DOE on additional areas of research that should be included in the FY 1997 program announcement. In view of the committee's recommendation that the DOE postpone the release of the program notice until later this year, the committee has decided to defer the consideration of this question to a future report in order to provide additional time for information gathering and deliberation. The committee plans to provide advice to the DOE on ways to identify and articulate its research needs in the program notice.
- Outreach to the research community: As noted earlier, the committee will consider ways in which DOE can improve outreach—both long and short term—to the research community and thereby improve the quality and relevance of the proposals submitted to the program.
- Program management: The committee will explore various models for managing the EMSP, drawing on the experiences of other federal and nonfederal institutions that manage "mission-linked" research programs. Such models might include field management with procurement authority, centers of excellence, consortia, and other cooperative arrangements. Additionally, the committee will consider the usefulness of advisory committees to keep the program focused and relevant to the long-term needs of the cleanup mission. The committee will consider the role of program management for ensuring that the program's research portfolio reflects an appropriate balance of problems, approaches, and levels of risk.
- Proposal evaluation: The committee will consider ways in which DOE can improve its evaluation of proposals through "scientific merit" and "mission relevance" reviews in order to identify projects that are likely to provide the greatest long-term payoffs to the cleanup mission.

---

[12]Communication from Dr. Carol Henry, Associate Deputy Assistant Secretary for Science and Technology, DOE.

Building an Effective EMSP: Final Assessment

FUTURE ACTIVITIES OF THE COMMITTEE

ER and EM face a formidable challenge in structuring and managing the EMSP to attract the best researchers and research ideas and to capitalize on the existing research infrastructure of the nation. This initial assessment of the EMSP has identified several major issues relevant to this challenge that will require the future attention of the committee. To examine these issues further, two panels of the committee will be established: the Panel on Science will focus on the science needs of the program, and the Panel on Management will evaluate the management structure and process.

The Panel on Science will obtain information on EM research needs and the basic research activities of other DOE programs and federal agencies in order to produce a report that addresses the following questions:

- How can science needs most effectively feed into the development of the EMSP research agenda?
- How can the program be structured to take advantage of research efforts and capabilities in other relevant DOE programs and federal agencies?
- How can the program be structured to broaden the community of researchers that can be called upon to address environmental problems?
- What areas of basic research are likely to provide the best payoffs for EM cleanup efforts over the next few decades?
- What additional areas of research should be included in future program notices as program evolves?

The Panel on Management will examine research program management and assessment in government and industry in order to produce a report that addresses the following questions:

- How can DOE evaluate the quality of the basic research it supports and the impact of this research on its cleanup mission?
- How can DOE identify changing needs for basic research as the program evolves?
- How should the program be structured and operated to assist the DOE in overall reduction of cleanup costs, risks, waste generation, and time requirements?

- How can the program be structured to take advantage of the unique capabilities of U.S. universities and federal labs?

The committee plans to meet at least three more times in the summer and fall of 1996 to gather information, deliberate on the issues, and write reports. A future meeting will be dedicated to a workshop at which panel members will have an opportunity to obtain information from and to question a broad group of invited university, national laboratory, industry, DOE, and other federal agency staff on the issues articulated above. The panels will issue final reports in late 1996.

REFERENCES

Blush, S.M., and Heitman, T.H. 1995. Train Wreck Along the River of Money: An Evaluation of the Hanford Cleanup. Report to the U.S. Senate Committee on Energy and Natural Resources, March.

Gephart, R. E., and Lundgren, R. E. 1995. Hanford Tank Clean up: A Guide to Understanding the Technical Issues. PNL-10773. Richland, Washington: Pacific Northwest Laboratory.

National Research Council (NRC). 1993. Science, Technology, and the Federal Government: National Goals for a New Era. Washington, D.C.: National Academy Press.

National Research Council. 1994. Alternatives for Ground Water Cleanup. Washington, D.C.: National Academy Press. http://www.nap.edu/nap/bookstore/0309049946.html

National Research Council. 1995. Allocating Federal Funds for Science and Technology. Washington, D.C.: National Academy Press.

National Research Council. 1996. Improving the Environment: An Evaluation of DOE's Environmental Management Program. Washington, D.C.: National Academy Press.

National Science Board. 1996. Science and Engineering Indicators. NSB 96-21. Washington, D.C.: U. S. Government Printing Office.

Office of Science and Technology Policy (OSTP). 1994. Science in the National Interest. Washington, D.C.: U.S. Government Printing Office.

U.S. Department of Energy (DOE). 1990. Basic Research for Environmental Restoration. DOE/ER-0482T. Office of Energy Research.

U.S. Department of Energy. 1995a. Alternative Futures for the Department of Energy National Laboratories. Prepared by the Secretary of Energy Advisory Board (Galvin commission). Washington, D.C.

U.S. Department of Energy. 1995b. Estimating the Cold War Mortgage: The 1995 Baseline Environmental Management Report. DOE/EM-0232. Office of Environmental Management.
http://www.em.doe.gov/bemr/index.html

U.S. Department of Energy. 1995c. Closing the Circle on the Splitting of the Atom: The Environmental Legacy of Nuclear Weapons Production in the United States and What the Department of Energy Is Doing About It. Department of Energy Office of Environmental Management.
http://www.em.doe.gov/circle/index.html

U.S. Department of Energy. 1995d. Natural and Accelerated In-Situ Bioremediation Program. DOE/ER-0659T. Office of Energy Research, Office of Health and Environmental Research.
http://www.er.doe.gov/production/oher/nabir/contents.html

U.S. Department of Energy. 1995e. Contaminant Plumes Containment and Remediation Focus Area: Technology Summary. DOE/EM-0248. Office of Environmental Management Technology Development.
http://www.em.doe.gov/emnet5.html

U.S. Department of Energy. 1995f. Decontamination and Decommissioning Focus Area: Technology Summary. DOE/EM-0253. Office of Environmental Management Technology Development.
http://www.em.doe.gov/emnet5.html

U.S. Department of Energy. 1995g. Landfill Stabilization Focus Area: Technology Summary. DOE/EM-0251. Office of Environmental Management Technology Development.
http://www.em.doe.gov/emnet5.html

U.S. Department of Energy. 1995h. Mixed Waste Characterization, Treatment, and Disposal Focus Area: Technology Summary. DOE/EM-0252. Office of Environmental Management Technology Development.
http://www.em.doe.gov/emnet5.html

U.S. Department of Energy. 1995i. Radioactive Tank Waste Remediation Focus Area: Technology Summary. DOE/EM-0255. Office of Environmental Management Technology Development.
http://www.em.doe.gov/emnet5.html

Zorpette, G. 1996. Hanford's nuclear wasteland. Scientific American 274(5).

APPENDIX A

ENVIRONMENTAL MANAGEMENT SCIENCE PROGRAM

The Environmental Management Science Program (EMSP) was initiated at the direction of the U.S. Congress, as noted in the introduction to this report. On February 9, 1996, the EMSP was announced jointly by the Offices of Energy Research (ER) and Environmental Management (EM). The program announcement (Program Notice 96-10; see Appendix A) was published in the Federal Register and on the World Wide Web, and a similar notification was sent to the national laboratories. As indicated in the program announcement, the objectives of this basic science program are to

• provide scientific knowledge that will revolutionize technologies and cleanup approaches to significantly reduce future costs, schedules, and risks;

• "bridge the gap" between broad fundamental research that has wide-ranging applicability, such as that performed in DOE's Office of Energy Research, and needs-driven applied technology development, conducted in EM's Office of Science and Technology; and

• focus the nation's science infrastructure on critical DOE environmental management problems.

By the preproposal deadline of February 28, 1996, DOE had received 2,200 applications. The preproposals were reviewed by ER research program managers and EM staff to determine whether the projects involved medium– to long–term basic research and were responsive to one or more of the priorities identified in the program announcement. After this review, 775 applicants were encouraged to submit full proposals. By May 8, 1996, DOE received 810 full proposals, of which approximately 270 were received from DOE laboratories and 540 from outside the DOE system, including universities and private organizations. A large number of multi-investigator and multi-institution proposals were also received.

The committee understands[12] that proposal review is being carried out in a two-step process—the first to assess scientific "merit" and the second to assess program "relevance"—that is being managed jointly by ER program managers and EM staff. Merit review is being obtained through

---

[12]Information on the proposal review process was provided to the committee by EM and ER staff during its two information-gathering meetings.

the use of peer review panels, comprised of scientists from industry, national laboratories, and universities, organized along disciplinary lines (Table A–1)[13], consistent with normal ER practices. The panels will discuss each of the proposals before them, and the panelists will provide individual ratings of each proposal as *must fund*, *should fund*, or *do not fund*. Following the panel meetings, federal ER program managers will determine an overall rating for each proposal.[14]

All of the proposals receiving overall ratings of *must fund* for scientific merit will be put forward for relevance review. Additionally, the proposals that received a strong recommendation of *should fund* will be put forward for review in case additional funds are available. This review will be undertaken by a panel of EM program managers from DOE headquarters and field offices who are knowledgeable of EM's needs and priorities. Federal ER program managers will participate in these reviews. The relevance review is scheduled for July 9, 1996, in Washington, D.C.

In July 1996, the Director of the Office of Energy Research will make final decisions on the awards with the concurrence of the Deputy Assistant Secretary for Science and Technology, Office of Environmental Management. Award funds will be obligated by the end of FY 1996. Program administration will be provided through DOE's Idaho field office.

---

[13]The panel meetings were held on June 17-25, 1996, in the Washington, D.C., area.

[14]The panels were not constituted under the Federal Advisory Committee Act and therefore are prohibited from determining a consensus rating.

TABLE A-1 Panels Convened by ER for Merit Review of EMSP Proposals

Review Panel(s)[a]	Number of Proposals
Plant Science	35
Analytical Chemistry	105
Separations Science	75
Catalysis	25
Heavy Elements Chemistry	40
General Inorganic Chemistry	50
Geophysics	35
Geochemistry	35
Flow Modeling	40
Flow, Field, and Bio/Geochemistry	55
Engineering Science	35
Materials Science	70
Applied Mathematics	10
Health Science and Risk Assessment	40
Bioremediation	160
Total	810

[a]Multiple panels were convened for areas that received large numbers of proposals (e.g., bioremediation).

APPENDIX B

EMSP PROGRAM NOTICE

Office of Energy Research
Office of Environmental Management

Federal Register: February 9, 1996 (Volume 61, Number 28)
Notices: Pages 4975-4978
From the Federal Register Online via GPO Access [wais.access.gpo.gov]

Energy Research Financial Assistance Program Notice 96–10;
Environmental Management Science Program

AGENCY: Department of Energy (DOE).

ACTION: Notice inviting grant applications.

SUMMARY: The Offices of Energy Research (ER) and Environmental Management (EM), U.S. Department of Energy, hereby announce their interest in receiving grant applications for performance of innovative, fundamental research to support the management and disposal of DOE radioactive, hazardous chemical, and mixed wastes.

This basic research should contribute to environmental management and restoration actions that would decrease risk for the public and workers, provide opportunities for major cost reductions, reduce time required to achieve EM's mission goals, and, in general, should address problems that are considered intractable without new knowledge. This program is designed to inspire "breakthroughs" in areas critical to the EM mission through long–term research and will be managed in partnership with ER. ER's well–established procedures, as set forth in the Energy Research Merit Review System, as published in the Federal Register, March 11, 1991, Vol. 56, No. 47, pages 10244-10246, will be used for merit review of applications submitted in response to this notice.

DATES: Potential applicants are strongly encouraged to submit a brief preapplication. All preapplications, referencing Program Notice 96–10, should be received by DOE by 4:30 p.m. EST, February 28, 1996. A response discussing the potential program relevance of a formal application generally will be communicated to the applicant within 15 days of receipt. The deadline for receipt of formal applications is 4:30 p.m., EDT, May 8, 1996, in order to be accepted for merit review and to permit timely consideration for award in fiscal year 1996.

ADDRESSES: All preapplications, referencing Program Notice 96–10, should be sent to Ms. Bobbi Parra, Office of Health and Environmental Research, ER–74, U.S. Department of Energy, 19901 Germantown Road, Germantown, Maryland 20874-1290, 301-903-3316, fax 301-903-8519, or by the internet e-mail address bobbi.parra@oer.doe.gov.

After receiving notification from DOE concerning successful preapplications, applicants may prepare formal applications and send them to: U.S. Department of Energy, Office of Energy Research, Grants and Contracts Division, ER–64, 19901 Germantown Road, Germantown, Maryland 20874-1290, Attn: Program Notice 96–10. The above address for formal applications must also be used when submitting formal applications by U.S. Postal Service Express Mail, any commercial mail delivery service, or when hand carried by the applicant. Please note that notification of a successful preapplication is not indication that an award will be made in response to the formal application.

It is anticipated that up to $20,000,000 will be available for grant awards during FY 1996 that will enable innovative fundamental research contingent upon the availability of appropriated funds. Multiple–year funding of grant awards is expected and is also contingent upon the availability of funds. Award sizes are expected to be on the order of $100,000-$300,000 per year for total project costs for a typical three year grant.

FOR FURTHER INFORMATION CONTACT: Dr. Michelle Broido, Office of Health and Environmental Research, ER–74, Office of Energy Research, 19901 Germantown Road, Germantown, Maryland 20874-1290. Telephone: (301) 903-3281, or Dr. Carol Henry, Office of Science and Risk Policy, Office of Environmental Management, 1000 Independence Avenue S.W., Washington, DC 20585, Telephone: (202) 586-7150.

SUPPLEMENTARY INFORMATION: The Office of Environmental Management, in partnership with the Office of Energy Research, is initiating an Environmental Management Science Program to fulfill DOE's continuing commitment for the cleanup of DOE's environmental legacy. Funding to initiate this program was established in the Conference Report accompanying the FY 1996 Energy and Water Development Appropriation Bill.

Purpose

The need to build a stronger scientific basis for the Environmental Management effort has been established in a number of recent studies and reports. Among the important observations and recommendations made by the Galvin Commission ("Alternative Futures for the Department of Energy National Laboratories," February 1995) are the following:

> There is a particular need for long term, basic research in disciplines related to environmental cleanup. Adopting a science-based approach that includes supporting development of technologies and expertise could lead to both reduced cleanup costs and smaller environmental impacts at existing sites and to the development of a scientific foundation for advances in environmental technologies.

The objectives of the basic science program are to:

• Provide scientific knowledge that will revolutionize technologies and clean-up approaches to significantly reduce future costs, schedules, and risks; and
• "Bridge the Gap" between broad fundamental research that has wide-ranging applicability such as that performed in DOE's Office of Energy Research and needs-driven applied technology development that is conducted in EM's Office of Science and Technology; and
• Focus the Nation's science infrastructure on critical DOE environmental management problems.

Representative Research Areas

Basic research is solicited for areas of concern to the Department's environmental management programs including but not limited to: chemical characterization of wastes and contaminants on an atomic and molecular level; development of knowledge of the physical and chemical behavior of such species; physical and chemical basis for waste separations and treatment; characterization and modeling of multi-phase chemical systems in natural systems, waste tanks and process streams; and monitoring, controlling, and assessing these processes. Understanding the fate of contaminants already in the environment includes the identification of the biological and geochemical reactions that sequester or degrade contaminants; understanding colloids or complexes of associated contaminants; and quantifying the impacts of geologic heterogeneity on the effectiveness of various remediation strategies. Indirect characterization of the geological environment by geophysical techniques provides the basic structural information essential in planning and monitoring remedial actions. Also important are studies to characterize flow and reactive transport through fractured and porous rocks and soils, and to characterize the physiological, biochemical, and genetic mechanisms for the uptake, transport, and sequestering of inorganic ions and organic molecules related to the use of plants and microorganisms for the cleanup of hazardous wastes.

Advances in information and monitoring technologies will also allow evaluation of progress in addressing these problems and devising new solutions. In the future, the focus will be on increasing efficiency in terms of materials and energy use. Better means of monitoring and controlling present system operations will significantly improve process efficiency and reduce waste outputs.

Specific examples illustrating the general subject areas, above, are found in the background section of this document.

Applicants in this program are strongly encouraged to collaborate with researchers in industry and/or the DOE National Laboratories, when appropriate, and to incorporate cost sharing and/or consortia wherever feasible. Grant applications are encouraged from all disciplines.

Merit Review and Evaluation Criteria

Formal applications will be subjected to formal merit review (peer review) and will be evaluated against the following evaluation criteria codified at 10 CFR 605(d).

1. Scientific and/or Technical Merit of the Project
2. Appropriateness of the Proposed Method or Approach
3. Competency of Applicant's Personnel and Adequacy of Proposed Resources
4. Reasonableness and Appropriateness of the Proposed Budget.

Examples of the considerations associated with determining the scientific and/or technical merit of the project include, but are not limited to:

—Potential for addressing problems identified by DOE, with meaningful progress within the proposed time frame.

—Benefits and merits of an application e.g. public purpose, time savings, extent of applicability, cost and risk reduction.

DOE shall also consider, as part of the evaluation, program policy factors such as an appropriate balance among the program areas.

Note, external peer reviewers are selected with regard to both their scientific expertise and the absence of conflict-of-interest issues. Non-federal reviewers may be used, and submission of an application constitutes agreement that this is acceptable to the investigator(s) and the submitting institution.

Preapplications

The brief preapplication, in accordance with 10 CFR 600.10(d)(2), should consist of two to three pages of narrative describing the research objectives and methods of accomplishment together with a brief summary of the principal investigator's publication and research background. The preapplications will be reviewed relative to the scope and research needs of the DOE's Environmental Management Science Program by qualified DOE program managers from both ER and EM. Telephone and FAX numbers are required parts of the preapplication, and electronic mail addresses are desirable.

Information

Information about the development, submission of applications, eligibility, limitations, evaluation, the selection process, and other policies and procedures may be found in 10 CFR Part 605, and in the Application Guide for the Office of Energy Research Financial Assistance Program. The Application Guide is available from the U.S. Department of Energy, Office of Energy Research, ER–74, 19901 Germantown Road, Germantown, Maryland 20874-1290. Telephone requests may be made by calling (301) 903-3316. Electronic access to ER's Financial Assistance Application Guide is possible via the World Wide Web at: http://www.er.doe.gov/production/grants/grants.html.

Background

The justification for such a program is grounded in the long–term costs for the Environmental Management program estimated at $200-350 billion over 75 years; in 10 years at current budget projections, $60 billion will have been spent, with over two thirds of the program yet remaining. This is the largest legacy from the Cold War of any other Federal program, dwarfing the Department of Defense's DOD's legacy by ten–fold. The Office of Environmental Management is responsible for waste management and cleanup of DOE sites. The EM operations have been historically compliance–based and driven to meet established goals in the shortest time possible using either existing technologies or those that could be developed and demonstrated within a few years. The Office of Energy Research addresses fundamental, frequently long–term, research issues related to the many missions of the Department. The Environmental Management Science Program will use ER's experience in managing fundamental research to address the needs of technology breakthroughs in EM's programs.

This research agenda has been initiated for Fiscal Year 1996, along with a development process for a long term program within the Office of Environmental Management, with the objective of providing continuity in scientific knowledge that will revolutionize technologies and clean–up approaches for solving DOE's most complex environmental problems.

Specific examples of areas of interest for research under this solicitation are:

• Advanced characterization methods that accelerate treatment and immobilization of high–level wastes. Pretreatment and separation methods that lead to a significant reduction in the amount of immobilized high–level waste requiring long–term isolation. Innovative separations for solids and for liquids, needed to significantly reduce projected high–level waste volume.

• In–situ characterization of dense non–aqueous phase liquid to allow comparative risk assessments of alternative treatment methods. In situ immobilization of subsurface contaminants to reduce pump and treat costs. Permeable in situ treatment barriers and factors governing in situ treatment processes to replace unsatisfactory, extant alternatives for treatment of large plumes. Degradation and extraction methods for radioactive and hazardous contaminants from soil/water. Dissolution of water–soluble sludge; washing of water soluble sludge, with recovery of cesium, strontium, technetium.

• Characterization of heterogeneous wastes needed to optimize decontamination and decommissioning recycling alternatives. Surface stabilization to reduce the ultimate waste volume and to enhance recycling. Selective and non–selective removal of contaminants from surfaces or bulk materials. Recycling of valuable commodities into general commerce.

• Non–destructive and in situ characterization methods to characterize the hazard of landfills. Innovative immobilization and transformation concepts that significantly reduce the cost of remediation. Ex–situ separation and treatment concepts to rapidly and safely destroy or immobilize landfill constituents.

• Emission–free destruction of organic wastes. Off–gas treatment that eliminates emissions in the environment that exceed Environmental Protection Agency requirements. Non–thermal treatment concepts for mixed waste. Bioremediation, enzymatic reactions, enzyme redesign, genetic engineering, microbial gene sequencing.

• Plutonium behavior in mixed matrices. Long–term monitoring concepts for plutonium.

• New concepts for waste stabilization of spent nuclear fuel. Long–term monitoring and performance assessment of spent nuclear fuel. Physics and chemistry of radionuclides in mixed matrices.

• Specialized waste forms. Performance assessment concepts for nuclear waste disposal.

• Ecology. Comprehensive understanding of the flow and use of materials and energy in our environmental system and the implications of those flows with respect to the environment. Ecosystem restoration and management; conduct monitoring, modeling, and process research to improve understanding of threatened and damaged ecosystems, technologies to restore the productivity and quality of these ecosystems.

• Biomarkers and sensors of exposure to contaminated media. Multi–site epidemiology studies. Effort to address current health concerns while continuing to conduct research that will promote a better future understanding of the relationship between exposure and health impacts.

The program will be competitive and offered to investigators in universities or other institutions of higher education, or other non–profit or for–profit organizations, non–Federal agencies or entities, or unaffiliated individuals. Apart from this notice, the program also will be offered to DOE national laboratories and other Federal laboratories, which will compete separately for appropriated funds. To ensure that the program is mission–oriented and that its achievements are recognized and used by EM, the Environmental Management Science Program will be closely integrated with EM's Technology Development Focus Areas and will also be closely coordinated with the Office of Energy Research to ensure use of broad–based fundamental research and development supported by that office.

Details of the programs of the Office of Environmental Management and the technologies currently under development or in use by the Environmental Management Program can be found on the World Wide Web at http://www.em.doe.gov and at the extensive links contained therein. These programs and technologies should be used as guidance when considering areas of research to be proposed.

The United States involvement in nuclear weapons development for the last 50 years has resulted in the development of a vast research, production, and testing network known as the nuclear weapons complex. The Department has begun the environmental remediation of the complex, encompassing radiological and nonradiological hazards, vast volumes of contaminated water and soil, and over 7,000 contaminated structures. The Department must characterize, treat, and dispose of hazardous and radioactive wastes that have been accumulating for more than 50 years at 120 sites in 36 states and territories. By 1995, the Department had spent about $23 billion in identifying and characterizing its waste, managing it, and assessing the remediation necessary for its sites and facilities. The Department estimates that the remedial actions at Department sites (not including groundwater cleanup, currently operating facilities and Naval facilities) could cost a total of $200–350 billion and take at least 75 years to complete. According to the estimates of the total program cost, 49% would go to waste management and 28% to environmental restoration, 10% to nuclear material and facility stabilization, and 5% to research and technology development with the remaining 8% for activities such as site security, transportation, and other landlord activities. The estimated life cycle costs over 75 years for the seven highest cost problem areas within the programs in descending order are as follows:

—Decommissioning
—High Level Waste
—Remedial Actions
—Low Level Waste
—Transuranic Waste
—Mixed Low Level Waste

—Spent Nuclear Fuel

Environmental Management is also responsible for conducting the program for waste minimization and pollution prevention for the Department. The variety and volume of the Department's current activities make this effort a challenge itself. In some cases, fundamental science questions will have to be addressed before a technology or process can be engineered. For example, improved understanding of the principles of pollutant transport in groundwater is required for important advancement in the development of effective groundwater–remediation technology. There is a need to involve more basic science researchers in the challenges of the Department's remediation effort.

References for Background Information on the Mission Responsibilities of the Office of Environmental Management

Note: World Wide Web locations of these documents are provided where possible. For those without access to the World Wide Web, hard copies of these references may be obtained by writing Dr. Carol Henry at the address listed in the contacts section.

DOE. 1995. Closing the Circle on Splitting of the Atom: The Environmental Legacy of Nuclear Weapons Production in the United States and What the Department of Energy is Doing About It. U.S. Department of Energy, Office of Environmental Management, Office of Strategic Planning and Analysis, Washington, DC.
http://www.em.doe.gov/circle/index.html

DOE. 1995. Estimating the Cold War Mortgage: The 1995 Baseline Environmental Management Report. Volume I, March 1995. U.S. Department of Energy, Office of Environmental Management, Washington, DC. http://www.em.doe.gov/bemr/index.html

DOE. 1995. Environmental Management 1995: Progress and Plans of the Environmental Management Program. The U.S. Department of Energy, Office of Environmental Management, Washington, DC. http://www.em.doe.gov/em95/index.html

DOE. 1995. Risks and the Risk Debate: Searching for Common Ground "The First Step". The U.S. Department of Energy, Office of Environmental Management, Washington, DC.
http://raleigh.dis.anl.gov:81/cgi–bin/dispdoc—return.pl?rrd+1

DOE. 1995. Technology Summary Reports, June 1995 (Rainbow Books) http://www.em.doe.gov/emnet5.html

DOE. 1995. Office of Science and Technology EM–50. http://www.em.doe.gov/emnet5.html

National Academy of Sciences. Allocating Federal Funds for Science and Technology. 1995. National Academy Press, Washington, DC. http://www.nas.edu/nap/online/fedfunds/

National Commission on Superfund Members. Final Consensus Report of the National Commission on Superfund. March 1994. Keystone Center and the Environmental Law Center of Vermont Law School. N/A

National Environmental Technology Strategy. Bridge to a Sustainable Future. April 1995. National Science and Technology Council, Washington, DC. http://iridium.nttc.edu/env/envstrat.txt

National Research Council. Improving the Environment: An Evaluation of DOE's Environmental Management Program. 1995. National Academy Press, Washington, DC. N/A

Secretary of Energy Advisory Board. Alternative Futures for the Department of Energy National Laboratories. February 1995. Task Force on alternative Futures for the Department of Energy National Laboratories, Washington, DC.
http://www.doe.gov/html/doe/whatsnew/galvin/tf–rpt.html

U.S. Congress, Office of Technology Assessment. Complex Cleanup: The Environmental Legacy of Nuclear Weapons Production, February 1991. U.S. Government Printing Office, Washington, DC. N/A

The Catalog of Federal Domestic Assistance Number for this program is 81.049, and the solicitation control number is ERFAP 10 CFR Part 605.

Issued in Washington, DC January 31, 1995. John Rodney Clark, Associate Director for Resource Management, Office of Energy Research. [FR Doc. 96–2877 Filed 2–8–96; 8:45 am] BILLING CODE 6450–01–P

APPENDIX C

MEETING AGENDAS

MEETING 1

<u>Saturday, May 11</u>

7:30–10:30   **Executive Session**

                             Open Session

11:00   Environmental Management Science Program/   *Carol Henry*
         Background and History   DOE

11:20   Environmental Management Science Program/ *MichelleBroido*
         Current Process   DOE

11:40   Questions and Discussions

12:00   Lunch

1:00   Questions and Discussions, continued

2:00   Panel Discussion on EM Science Program/   *Sally Benson*
         Opportunities and Challenges   *Gregory Choppin*
                                                  Donald J. DePaolo
                                                      A.J. Francis
                                                      Remy Hennet
                                                   Michael Knotek
                                                  Terrence Surles

3:45   Break

4:00–5:30   **Executive Session**

<u>Sunday, May 12</u>

8:00–1:30   **Executive Session**

42

MEETING 2

<u>Saturday, June 15</u>

7:45–11:15	**Executive Session**	
	Open Session	
11:30	Welcome; progress report and plan for the meeting	*Chair*
11:35	Reflections on the first committee meeting	*Carol Henry* *Ari Patrinos*
12:15	Working Lunch	
1:20	EM Science: Challenges and Opportunities	*Judy Bostock*
2:00	Planning for the Science and Management Workshops	*All*

	Objectives
	    Structure and Organization
	    Products
	    Schedules and Locations	

3:30	Break	
3:45	Breakout into Science/Management Groups to Develop Preliminary Workshop Agendas	
5:00	Breakout Group Reports	*Ahearne* *Silver*
5:30	Appointment of Subcommittees	
6:00	Adjourn	

44   *Building an Effective EM Science Program*

Sunday, June 16

   7:30   **Executive Session**

   1:00   Adjourn

APPENDIX D

BIOGRAPHICAL SKETCHES OF COMMITTEE MEMBERS AND CONSULTANTS

AHEARNE, John F.—Dr. Ahearne received his B.S. and M.S. degrees from Cornell University and his Ph.D. in plasma physics from Princeton University. He has served as commissioner and chairman of the U.S. Nuclear Regulatory Commission, system analyst for the White House Energy Office, Deputy Assistant Secretary for Energy, and Principal Deputy Assistant Secretary for Defense. He currently is the director of the Sigma Xi Center for Sigma Xi, The Scientific Research Society, and a lecturer in public policy at Duke University. Dr. Ahearne is a member of the Department of Energy's Environmental Management Advisory Board and the National Research Council's Board on Radioactive Waste Management, and has served on a number of the National Research Council's committees examining issues in risk assessment. His professional interests are reactor safety, energy issues, resource allocation, and public policy management. He is a fellow of the American Physics Society, American Association for the Advancement of Science, and American Academy of Arts and Sciences. He is a member of Sigma Xi, the Society for Risk Analysis, the American Nuclear Society, and the National Academy of Engineers.

ARNETT, Edward M.—Dr. Arnett earned a B.A., M.S., and Ph.D. in chemistry from the University of Pennsylvania. He is professor emeritus of chemistry at Duke University and has held prior professorships at the University of Pittsburgh and Western Maryland College. His expertise is in organic and physical organic chemistry. He is a Guggenheim fellow and has received numerous awards, including most recently the Arthur C. Cope Scholar Award and the American Institute of Chemists Distinguished North Carolina Chemist Award. Dr. Arnett is a member of the National Academy of Sciences.

AUERBACH, Stanley I.—Dr. Auerbach earned his B.S. and M.S. from the University of Illinois, and his Ph.D. in zoology from Northwestern University. Dr. Auerbach retired as director of the Environmental Sciences Division at Oak Ridge National Laboratory in 1990. His research interests include radiation ecology ecosystem analysis and radioactive waste cycling in terrestrial ecosystems. Dr. Auerbach's former academic positions include lecturer and adjunct professor at the University of Tennessee and visiting professor at the University of Georgia. He has served on or chaired several National Research Council committees, boards, and commissions since 1961. He is a member of the American Institute for Biological Science, American Association for the Advancement of Science, Ecological Society of America, British Ecological Society, International Union of Radioecologists, and Health Physics Society.

BOUWER, Edward J.—Dr. Bouwer received his B.S.C.E. from Arizona State University in civil engineering and his M.S. and Ph.D. in environmental engineering and science from Stanford University. He is currently a professor of environmental engineering at Johns Hopkins University. His research interests include biodegradation of hazardous organic chemicals in the subsurface, biofilm kinetics, water and waste treatment processes, and transport and fate of bacteria in porous media. He serves on the board of directors for the Association of Environmental Engineering Professors and on the editorial boards for The *Journal of Contaminant Hydrology* and *Biodegradation*. He has served on three past National Research Council committees.

BRAUMAN, John I.—Dr. Brauman earned a B.S. from the Massachusetts Institute of Technology and a Ph.D. in chemistry from the University of California at Berkeley. Dr. Brauman is the J.G. Jackson–C.J. Wood Professor of Chemistry at Stanford University. He began his career at Stanford University in 1963 as an assistant professor. His research interests include physical and organic chemistry, gas phase ionic reactions, electron photodetachment spectroscopy, and reaction mechanisms. He is the recipient of many awards from the American Chemical Society, including the Award in Pure Chemistry, the James Flack Norris Award in Physical Organic Chemistry, and the Arthur C. Cope Scholar Award. Dr. Brauman is a Guggenheim fellow and an honorary fellow of the California Academy of Sciences; he is a member of the National Academy of Sciences, the American Academy of Arts and Sciences, and the American Chemical Society. He has served on several National Research Council committees.

HARLEY, Naomi H.—Dr. Harley holds a B.E. in electrical engineering from the Cooper Union and an APC in management from the New York University Graduate Business School. She received an M.E. in nuclear engineering and a Ph.D. in radiological physics from New York University. Dr. Harley is a research professor of environmental medicine at the New York University Medical Center where she also serves on the Medical Isotopes Committee. Her expertise is in radiation carcinogenesis, and her major research interests include measurement of inhaled or ingested radionuclides, modeling of their fate within the human body, and the calculation of the detailed radiation dose to the cells specific to carcinogenesis. She is a member of the National Council on Radiation Protection and Measurements and an adviser to the U.S. Delegation of the United Nations Committee on the Effects of Atomic Radiation. Dr. Harley is a member of the editorial board of *Environment International*, and a fellow of the Health Physics Society; she holds three patents at New York University for radiation detection devices.

LOVLEY, Derek R.—Dr. Lovley received a B.A. in biological sciences from the University of Connecticut, an M.A. from Clark University, and a Ph.D. in microbiology from Michigan State University. He is a professor of microbiology at the University of Massachusetts, Amherst. His research interests comprise the physiology and ecology of novel anaerobic microorganisms, molecular analysis of anaerobic microbial communities, and bioremediation of metal and organic contamination. He is an associate editor for *Anaerobe* and is on the editorial boards of *Applied and Environmental Microbiology*, *Microbial Ecology*, and *FEMS Microbiology Ecology*.

MANNELLA, Gene G.—Dr. Mannella earned a B.S. from Case Institute of Technology and a Ph.D. in chemical engineering from Rensselaer Polytechnic Institute. He retired in 1994 as senior vice president of business operations, at the Gas Research Institute, headquartered in Chicago. He has also served as director of the Washington office of the Electric Power Research Institute, vice-president and general manager of Mechanical Technology, Inc., and senior vice-president at the Institute of Gas Technology. Dr. Mannella has held several positions in government agencies including the National Aeronautics and Space Administration,

Department of Transportation, and Energy Research and Development Administration (predecessor to the Department of Energy). He has authored numerous technical papers and served on several committees and boards including the Washington Coal Club.

NOONAN, Norine E.—Dr. Noonan received her B.A. from the University of Vermont, summa cum laude, in zoology/chemistry, and her M.A. and Ph.D. degrees in cell biology and biochemistry from Princeton University. She is vice president for research and dean of the Graduate School at Florida Institute of Technology in Melbourne. Prior to joining Florida Tech in October 1992, Dr. Noonan was chief of the Science and Space Programs Branch of the Energy and Science Division, Office of Management and Budget. In this capacity, she was responsible for the legislative programs and combined budgets. Before becoming branch chief, Dr. Noonan was senior budget and program analyst for the branch for four years. She was an American Chemical Society Congressional Science Fellow for the United States Senate Committee on Commerce, Science, and Transportation; a research associate professor of biochemistry at Georgetown University School of Medicine; an expert consultant for the Subcommittee on Science Research and Technology; and associate professor of physiological sciences at the University of Florida, College of Veterinary Medicine. Dr. Noonan is a member of the American Association for the Advancement of Science and is also a member of the American Society for Cell Biology, Sigma Xi, and Phi Beta Kappa.

SILVER, Leon T.—Dr. Silver earned a B.S. in civil engineering from the University of Colorado, an M.S. in geology from the University of New Mexico, and a Ph.D. from the California Institute of Technology. He is the W.M. Keck Foundation Professor for Resource Geology at the California Institute of Technology (CalTech) and his expertise is in petrology and geochemistry. Dr. Silver was a public works officer in the U.S. Naval Civil Engineer Corps from 1945 to 1946 and held several positions at the United States Geological Survey before he joined CalTech. He has served on numerous National Research Council committees, including his current membership of the Commission on Physical Sciences, Mathematics, and Applications. Dr. Silver is a member of the National Academy of Sciences.

CONSULTANTS

CHOPPIN, Gregory R.—Dr. Choppin received a B.S. in chemistry from Loyola University, New Orleans, and a Ph.D. from the University of Texas, Austin. He is currently the R.O. Lawton Distinguished Professor of Chemistry at Florida State University. His research interests involve the chemistry of the f–elements, the separation science of the f–elements, and concentrated electrolyte solutions. During a postdoctoral period at the Lawrence Radiation Laboratory, University of California, Berkeley, he participated in the discovery of mendelevium, element 101. His research and educational activities have been recognized by the American Chemical Society Award in Nuclear Chemistry, the Southern Chemist Award of the American Chemical Society, the Manufacturing Chemist Award in Chemical Education, a Presidential Citation Award of the American Nuclear Society, and honorary D.Sc. degrees from Loyola University and the Chalmers University of Technology (Sweden).

DePAOLO, Donald J.—Dr. DePaolo earned a B.S. with honors from the State University of New York, Binghamton, and a Ph.D. from the California Institute of Technology. He is professor of geochemistry and director of the Center for Isotope Geochemistry at the University of California, Berkeley. Prior to arriving at Berkeley in 1988, Dr. DePaolo held a professorship at the University of California, Los Angeles. He is a recipient of the F.W. Clarke Medal of the Geochemical Society, the J.B. MacElwane Award of the Geophysical Union, and the Mineralogical Society of America Award. He is a member of the National Academy of Sciences.

HORNBERGER, George M.—Dr. Hornberger received an undergraduate degree in civil engineering, but subsequently trained as a hydrologist at Stanford University, where he was awarded a Ph.D. in 1970. Dr. Hornberger is currently the Ernest H. Ern Professor of Environmental Sciences at the University of Virginia. He joined the University of Virginia's Environmental Sciences Department in 1970 and served as department chairman from 1979 to 1984. Dr. Hornberger has been the recipient of numerous awards, including election to the first group of fellows of the Association for Women in Science. He was cited for "exemplary commitment to the achievement of equity for women in science and technology." Dr. Hornberger received the John Wesley Powell Award from the U.S. Geological Survey and is also a member of the American Geophysical Union. He is the editor of *Water Resources Research*, the nation's premier journal for publications in the hydrological sciences. He was elected to the National Academy of Engineering in 1996.

APPENDIX E

ACRONYMS

DOD	United States Department of Defense
DOE	United States Department of Energy
DOE–EM (EM)	United States Department of Energy, Office of Environmental Management
DOE–EMSP (EMSP)	United States Department of Energy, Environmental Management Science Program
DOE–ER (ER)	United States Department of Energy, Office of Energy Research
EM–50	United States Department of Energy, Office of Environmental Management, Office of Science and Technology
EPA	United States Environmental Protection Agency
FY	Fiscal Year
GPO	Government Printing Office
H. R.	House of Representatives Bill
NABIR	Natural and Accelerated In–Situ Bioremediation Program
NAS	National Academy of Sciences
NRC	National Research Council
NSF	National Science Foundation
OSTP	Office of Science and Technology Policy
PNL	Pacific Northwest Laboratory
R&D	Research and Development
USGS	United States Geological Survey

APPENDIX G

LETTER REPORT

NATIONAL RESEARCH COUNCIL

VIRTUAL COMMISSION ON ENVIRONMENTAL MANAGEMENT SCIENCE

2101 Constitution Avenue Washington, D.C. 20418

Executive Office   202/334-3066

October 8, 1996

Dr. Carol Henry
Office of Science and Risk Policy
U.S. Department of Energy
Washington, D.C. 20585

Dear Dr. Henry:

In response to your letter of August 9, 1996, the National Research Council's Committee on Building an Environmental Management Science Program offers this letter report on the fiscal year (FY) 1997 program announcement for the U.S. Department of Energy's (DOE's) Environmental Management Science Program (EMSP).

The committee has been charged to advise the Department on ways to improve the effectiveness of the EMSP. The statement of task for the committee's work is given in Attachment A. The committee members were selected to provide a balance of expertise and perspectives, including knowledge of and experience with the weapons complex and its clean-up challenges, and the proposal solicitation process related to basic research. A list of committee members is given in Attachment B.

The committee held its first meeting on May 11-12, 1996, and published the first of three reports, *Building an Effective Environmental Management Science Program: Initial Assessment*, in July.[1] The *Initial Assessment* report presents the committee's preliminary evaluation of the EMSP, findings regarding the FY 1996 proposal competition, and recommendations for the FY 1997 program announcement. The committee specifically recommended that the Department postpone the release of the FY 1997 program announcement to allow time to identify and incorporate lessons learned from the FY 1996 program competition and to determine how the program should be structured and managed. The committee also noted that the FY 1997 competition likely will have a major role in shaping the program and ensuring its future success.

Reasons for Writing this Letter Report

Your letter of August 9, 1996 (Attachment C), requests that the committee provide additional advice to the Department regarding the content of the FY 1997 program announcement, and, in particular, advice on research needs. This letter is meant to address this request. This letter reflects a consensus of the committee and has been reviewed in accordance with the procedures of the National Research Council. This letter does not take the place of the committee's final report, which will be completed by the end of the year, but rather is intended to

---

[1] National Research Council. 1996. *Building an Effective Environmental Management Science Program: Initial Assessment.* Washington, D.C.: National Academy Press. This report is available on the World Wide Web at the following address: http://www.nap.edu/readingroom/ books/envmanage/index.html.

Dr. Carol Henry
October 8, 1996
page 2

provide the Department with more timely advice to avoid an unnecessary delay in the release of the FY 1997 program announcement.

Information Sources for this Letter

As a first step in its deliberations that led to this letter, the committee reviewed the results of the FY 1996 proposal competition to assess the effectiveness of the FY 1996 program announcement. To this end, the committee asked for—and received—from the Department the following data on the successful proposals for the FY 1996 proposal competition:

- proposal titles and names of principal investigators (PIs);
- their institutional affiliations;
- award amounts; and
- scientific field of proposal (e.g., geoscience) and area of potential impact of the proposed research (e.g., contaminant plume treatment).

Because the information provided to the committee lists only the principal investigator of each project, evidence of collaborations between individuals or institutions is lacking. Thus, while collaborations may exist, the committee was not able to determine them from the information provided. The committee is seeking additional information on collaborations for its final report.

The information listed above was provided to the committee the day before its fourth meeting, which was held on August 21-22, 1996. The committee concluded that it did not have enough time or information to assess conclusively the overall success of the FY 1996 program and the effectiveness of the FY 1996 program announcement. However, the committee did have first-hand information on the makeup and operation of one of the review panels and was able to confirm the overall quality of the proposals, the review process, and the review panelists. The committee intends to provide additional comments on the success of the FY 1996 competition in its final report.

The committee also requested and was provided with a list of the titles of unsuccessful projects. This information allowed the committee to inform itself generally on the nature of proposed projects, but the titles themselves did not provide the members with enough information to make an effective assessment of the quality of the research or proposers. The committee does note, however, that the titles indicate that the Department received proposals in a wide range of research areas listed in the FY 1996 program announcement.

The committee also received copies of the guidelines that were given to the merit and relevance review panelists by the offices of Environmental Management (EM) and Energy

Dr. Carol Henry
October 8, 1996
page 3

Research (ER).[2] The committee requested—but did not receive—the names of the merit and relevance review panelists, and the final ratings of the proposals. This information is considered confidential by the Department.

The committee expects to receive additional information on the FY 1996 proposal competition at a later date. This information includes abstracts of the successful projects, biographical sketches of investigators who received funding, and, for successful investigators, a list of recent, current, and pending research support. The committee is especially interested in the number of successful investigators with recent, current, or pending Department support relative to the total number of investigators. The committee believes that this information will help it to assess whether the EMSP was successful in attracting high-quality researchers and innovative proposals to the program.

Recommendations for the FY 1997 Program Announcement

The FY 1996 program announcement provided a fairly complete description of the EMSP. The announcement included a statement of purpose, a list of research needs, a brief description of the criteria used for proposal review and selection, a schedule for proposal submission and review, and a financial plan. The committee believes that such a self-contained announcement is helpful to the research community because it provides most of the information needed to prepare a competitive proposal in a single, readily accessible package. The committee recommends that the Department use the same approach in developing the FY 1997 program announcement, and it offers suggestions below on the following elements of the announcement:

- criteria for proposal review and selection;
- research areas;
- proposal format;
- program schedule;
- review process; and
- financial plan.

1. *Criteria for proposal review and selection.* The FY 1997 program announcement should be explicit about what criteria will be used to select proposals for funding. The committee recommends that the Department utilize the following criteria:

---

[2] The EMSP used a two-phased approach in reviewing proposals, one review for scientific merit and one for relevance to EM clean up needs. The relevance review, conducted by EM, is essentially a federal review by EM program managers. The merit review was conducted by panels of scientists and engineers convened by ER. For additional details, see the committee's first report.

Dr. Carol Henry
October 8, 1996
page 4

• *Focus on basic research.* The purpose of the EMSP is to foster basic research that will contribute to long-term clean up of the weapons complex. The focus on basic research should be articulated clearly in the program announcement.

• *Scientific merit.* As noted in the committee's *Initial Assessment* report, merit should be the primary criterion for proposal selection. Using scientific merit as a first screen will help ensure that only high-quality proposals are supported by the program, and it will help keep the focus of the program on basic research.

• *Relevance to mission.* Also as noted in the *Initial Assessment* report, research should be broadly relevant to EM's clean up mission. That is, the basic research supported in the EMSP should address the phenomena and processes that underpin EM's clean up problems. The proposal need not demonstrate knowledge of problems at weapons complex sites to be useful to the clean up mission in the long term, and such knowledge should not be required in the proposal.

• *A demonstrated record of research accomplishment.* As noted in the committee's *Initial Assessment* report, the EMSP should aim to attract outstanding researchers to work on EM's problems.

• *The project must be able to demonstrate progress (but not necessarily completion) in the 3-year time period.* As mandated by Congress, one purpose of the program is to "stimulate the required basic research," which may require longer–term commitments beyond 3 years. The committee recognizes, of course, that even long-term projects may yield important "deliverables" over much shorter time frames.

• *Training opportunities.* In its *Initial Assessment* report, the committee commented on the need to build a "committed cadre" of researchers for the EMSP. The committee believes that graduate student training is an effective mechanism for building a community of researchers knowledgeable of EM's problems and responsive to EM's research needs. The program announcement should encourage (but not require) graduate student involvement in research proposals submitted to the program.

2. *Research areas.* In the committee's original statement of task (Attachment A) and your letter requesting this report (Attachment C), the committee was asked to identify additional areas of research that should be included in the EMSP. In its deliberations on this issue, the committee has concluded that the EMSP is more likely to attract innovative proposals from creative researchers if the focus of the program announcement is shifted away from a statement of suggested solutions (i.e., *research areas*), as was provided in the FY 1996 program announcement, to a statement of EM's *problems* that require basic research. As an aid to researchers, the Department also may wish to include in its program announcement examples of

Dr. Carol Henry
October 8, 1996
page 5

innovative proposals that were submitted by researchers in the FY 1996 proposal competition. The committee believes that this approach would encourage researchers who are not knowledgeable of EM's clean up problems to apply their expertise and suggest solutions that may not have occurred to the authors of the program announcement.

It is beyond the experience and the expertise of the committee to provide a list of EM problems that should be included in the FY 1997 program announcement. In its *Initial Assessment* report, the committee recommended that "concise technical summaries" of clean-up problems be prepared by the Department. The committee reaffirms the importance of these summaries and recommends that they be prepared forthwith. Such summaries should include examples of the types of problems that exist at specific sites as well as more generic problems that apply across sites, such as ground water contamination.

In formulating this problem list, the committee encourages the Department to broaden the solicitation to include problems related to risk, quantitative methodologies, and health assessment. As noted in two recent National Research Council reports,[3] relevant research on risk would be especially valuable for prioritizing clean up efforts and allocating limited resources. Indeed, a risk-based approach is currently being used by the Department to help identify and rank the important problems and prioritize clean up (DOE, 1995).[4] Currently, there is much scientific uncertainty about the very existence of risk to human health at the low levels projected for the end stages of the clean-up effort. To establish standards and measures of progress, substantial improvement in the scientific state-of-art is needed. The EMSP could contribute further to the understanding of risk and risk-based approaches to priority setting. Accordingly, the committee recommends that the program announcement be expanded to include risk as it relates to the clean-up program, both now and in the future.

3. *Proposal format.* To emphasize important information that is required in a proposal, the committee recommends that the Department specify a format for proposals that incorporates the elements shown in Appendix D. A standard format would be a major aid to reviewers in assessing and comparing proposals.

---

[3] National Research Council. 1995. *Improving the Environment: An Evaluation of DOE's Environmental Management Program.* Washington, D.C.: National Academy Press; National Research Council. 1994. *Building Consensus Through Risk Assessment and Management of the Department of Energy's Environmental Remediation Program,* Washington, D.C.: National Academy Press.
[4] U.S. Department of Energy, Office of Environmental Management. *Risks and the Risk Debate: Searching for Common Ground "The First Step".* 1995.

Dr. Carol Henry
October 8, 1996
page 6

4. *Program schedule.* The accelerated FY 1996 competition schedule presented a significant challenge to researchers, review panelists, and EM and ER program managers.[5] The compressed schedule in the first round gave researchers little time to educate themselves on EM clean-up problems, to develop proposals, or to establish new collaborations. For program managers and review panelists, the tight schedule placed severe pressures on the preproposal selection process and final proposal reviews. The committee believes that more time should be allowed in the FY 1997 program competition to alleviate these pressures. To this end, the committee recommends that the Department provide researchers with at least one month to prepare preproposals and two months to prepare full proposals.

The committee believes that the FY 1997 review process also would be improved by giving review panelists more time to examine proposals. In the FY 1996 competition, most merit panelists received proposals to review only two weeks prior to the panel meetings, and most relevance panelists did not receive proposals in advance of their meetings. The committee believes that the panelists will do a better job of evaluating these proposals if they are given more time to review them prior to their panel meetings.

5. *Review process.* The committee recommends that the FY 1997 program announcement provide a clear description of the process that will be used to review proposals and select awards. The committee offers the following comments and recommendations for the Department's consideration.

The collaborative management efforts between ER and EM have been very successful to date, and the committee urges continued interactions and open communications between staff in these offices in the FY 1997 program competition. The committee reaffirms its endorsement (from the *Initial Assessment* report) of the two-phase review process used in the FY 1996 competition that first evaluates the scientific and technical merit of the proposals and then examines more closely the relevance of the proposed work to the clean-up mission. The committee believes that this two-phase review should continue in FY 1997 and that it should continue to be managed as a partnership between ER and EM.

The committee further recommends that the Department retain, to the extent possible, continuity in merit and relevance review panels to take advantage of the experience gained in the FY 1996 competition. Additionally, the committee recommends that the Department's preproposal screening process involve, to the extent possible, members of the merit and relevance panels. The involvement of the research and clean-up communities in preproposal review will

---

[5] As noted in the *Initial Assessment* report, the program announcement was published in the Federal Register on February 9, 1996. The deadline for submission of preproposals was February 28, and full proposals were due by May 8. The proposals were reviewed in July and awards were announced in August. Awards were made in September.

Dr. Carol Henry
October 8, 1996
page 7

strengthen the preselection process, which in the FY 1996 competition eliminated roughly two-thirds of the preproposals.

In the FY 1996 EMSP competition, the merit review panels convened by ER[6] were constituted as non-FACA[7] committees. In this capacity, the panelists were allowed to discuss the proposals and to provide ER program managers with individual scores on each proposal. The panels were not allowed to reach consensus, nor were they allowed to provide ER program managers with a rank ordering of the proposals considered by each panel. Further, the names of the panelists were kept confidential to the proposers and the research community at large, including this committee.[8]

As noted in its *Initial Assessment* report, building credibility in the research community is a singular challenge for the EMSP. The committee believes that such credibility is less likely to be achieved when the review process has the appearance of a "black box" into which proposals are fed and out of which funding decisions emerge. To achieve more transparency in the process—and to provide for a higher quality of merit review by allowing panelists to reach consensus on proposal scoring and ranking—the committee strongly recommends that ER follow established practices of other federal agencies with basic research programs, such as the National Institutes of Health and the National Science Foundation, by constituting the FY 1997 review panels as FACA committees. The committee further recommends that ER announce its intention to follow the FACA process for merit review in the FY 1997 program announcement.

The committee recognizes that this recommendation may be difficult to implement given DOE's history with the FACA process and ER's current practice with respect to FACA review panels.[9] Nevertheless, the committee offers this recommendation because it believes that merit review panels constituted under FACA will improve significantly the quality and credibility of the review process.

6. *Financial plan.* In its *Initial Assessment* report, the committee recommended that "successful proposals should be funded fully 'up front' to help ensure the stability and continuity of the research projects and to establish a solid foundation on which a stable, long-term program can be built." The committee believes that the full-funding of proposals is essential for

---

[6] See Appendix A of the committee's *Initial Assessment* report for a description of the EMSP proposal review process.

[7] FACA denotes the Federal Advisory Committee Act.

[8] The EM relevance review panels were comprised of federal program managers and thus do not fall under the federal act. The committee understands that EM decided to keep the names of the relevance review panelists confidential to be consistent with the ER review process.

[9] At present, ER does not convene its review panels under FACA.

Dr. Carol Henry
October 8, 1996
page 8

establishing credibility for the program in the research community and, therefore, is an important factor in attracting high-quality researchers and proposals.

In FY 1996, the Department was able to provide full funding for proposals submitted by non-DOE performers (i.e., proposals from universities, industry, and non-profit research performers), but DOE was not able to provide full funding for proposals from national laboratory performers. The Department committed about $112 million in the FY 1996 competition. A total of $43 million was provided out of FY 1996 funds to provide full funding for the 3-year non-DOE projects, and $4 million was provided to national laboratory projects. The remainder, about $65 million, will be provided to national laboratory projects out of future-year program funds (i.e., FY 1997, FY 1998, and FY 1999 funds). The committee believes that this "mortgage" represents a significant challenge to the future viability of the program. In FY 1997, for example, $23 of the $50 million allocated to this program[10] already has been committed to funding FY 1996 projects at national laboratories.

The committee has reviewed the future-year commitments from the FY 1996 awards to the national laboratories and has concluded that, on the current path, considerably fewer new or competitive renewal awards will be made in future years unless significantly more funding is made available. Attachment E provides two simple scenarios for future funding that illustrate the committee's concerns regarding the "mortgage" and balance of funding for universities and national laboratories. Table E.1 shows that if the current pattern of funding is continued, approximately $112 million in program funds will be required annually by FY 2000 to maintain current levels of funding for new or renewal projects. If funding is constrained to approximately $50 million per year—the amount of funding available to the program in FY 1996—then funds for new or competitive renewal projects will decrease by approximately 75 percent.

Recognizing that a serious funding problem may be developing, the committee strongly encourages the Department to explore mechanisms to provide full funding for successful national laboratory proposals for the FY 1997 proposal competition. This issue should be resolved, if possible, before the FY 1997 program announcement is released, because it will govern the amount of funding available to the program next year, and hence the number of new starts.[11] Additionally, the committee recommends that funding guidelines, but not dollar limits, be provided in the FY 1997 program announcement. Specific dollar limits may restrict potentially outstanding research proposals from being submitted, which in turn, could limit the development of effective academic-laboratory-industry partnerships.

---

[10] Conference Report on H.R. 3816, Energy and Water Development Appropriations Act, 1997 (Congressional Record, v. 142, no. 125, p. H10320).
[11] Fewer awards can be made if the Department provides full funding for all successful proposals next year.

Dr. Carol Henry
October 10, 1996
page 9

Announcements/Publication

The Department published the FY 1996 program announcement in the Federal Register and on its home page, and it sent notices by mail to approximately 200 universities. The committee encourages the Department to utilize these dissemination mechanisms to publicize the FY 1997 proposal competition, and it recommends that the Department explore additional mechanisms to make the research community more broadly aware of the FY 1997 proposal competition. To this end, the committee recommends the use of paid advertisements in professional journals such as *Science, Chemical and Engineering News,* and *EOS* to publicize this program in FY 1997.

Summary

The FY 1997 program announcement will have a major impact on the future direction and viability of the EMSP. Although the committee had previously recommended postponing the release of the FY 1997 program notice in its *Initial Assessment* report, the committee recognizes the urgency of the Department's request for advice on the content of the notice. With the suggestions for modifications to the FY 1997 program announcement provided in this letter, the committee now urges the Department to move forward expeditiously to release the program announcement as soon as possible.

Sincerely,

John Ahearne, Chair
Committee on Building an Environmental
Management Science Program

Attachments:
A. Statement of Task
B. Committee Membership
C. August 9, 1996 Letter from Carol Henry Requesting this Report
D. Sample Proposal Format
E. Funding Projections

ATTACHMENT A

Statement Of Task

The committee will produce two reports that address the science and management needs of the Department of Energy's (DOE's) Environmental Management (EM) Science Program. These reports will be produced in two separate activities as noted below.

ACTIVITY #1: FY97 RESEARCH PROGRAM

The committee will draw on the expertise of its members and other outside experts, the results of the 1996 DOE workshops on research needs, and previous NRC and federal government reports in order to address the following questions:

1. How can basic research be used to help DOE EM "to complete its mission successfully in the next few decades"?
2. How can a basic research program help add value to DOE EM's cleanup efforts?
3. What kinds of technical challenges would likely benefit from a program in basic research?
4. How can the research program take advantage of the unique capabilities of U.S. universities and federal labs?
5. How can the research program take advantage of research efforts and capabilities in other DOE programs and other federal agencies?
6. What, if any, additional areas of research should be included in the fiscal year (FY) 1997 program announcement as the DOE EM Science Program evolves?

The committee will not attempt to be comprehensive in addressing these questions, but, rather, its focus will be on providing guidance to DOE-EM for use in the FY97 program solicitation.

ACTIVITY #2: SCIENCE AND MANAGEMENT NEEDS

The committee will produce a final report that provides a more detailed assessment of the science and management needs of the EM Science Program. This report will address the following questions:

Science Needs

1. How can science needs most effectively feed into the development of the EM research agenda?
2. How can the research program be structured to take advantage of research efforts and capabilities in other DOE programs and other federal agencies? (The committee would revisit the issue from the first activity.)

Dr. Carol Henry
October 8, 1996
page A-2

3. How can the research program be structured to broaden the community of researchers that can be called upon to address environmental problems?

4. What areas of basic research are likely to provide the best payoffs for EM cleanup efforts over the next few decades?

5. What additional areas of research should be included in future program announcements as the DOE EM Science Program evolves? (The committee would revisit the issue from the first activity.)

Management Needs

1. How can the DOE evaluate the quality of the basic research it supports and the impact of this research on its cleanup mission?

2. How can DOE identify changing needs for basic research as the program evolves?

3. How should the program be structured and operated in order to assist the DOE in overall reduction of cleanup costs, risks, waste generation, and time requirements?

4. How can the program be structured take advantage of the unique capabilities of U.S. universities and federal labs? (The committee would revisit the issue from the first activity.)

Sponsor(s): Department of Energy
Date of Statement: 10/8/96
Date of Previous Statement: 7/15/96

ATTACHMENT B

Steering Committee on Building an Environmental Management Science Program

John F. Ahearne, *CHAIR*
Sigma Xi, The Scientific Research Society & Duke University

Edward M. Arnett
Duke University (emeritus)

Stanley I. Auerbach
Oak Ridge National Laboratory (retired)

Edward J. Bouwer
The Johns Hopkins University

John I. Brauman
Stanford University

Naomi H. Harley
New York University Medical Center

Harold Lewis
University of California, Santa Barbara (emeritus)

Derek R. Lovley
University of Massachusetts, Amherst

Alexander MacLachlan
DuPont (retired)

Gene G. Mannella
Gas Research Institute (retired)

Norine E. Noonan
Florida Institute of Technology

Jerome Sacks
National Institute of Statistical Sciences

Alfred P. Sattelberger
Los Alamos National Laboratory

Leon T. Silver
California Institute of Technology

COMMITTEE CONSULTANTS

Gregory R. Choppin
Florida State University

Donald J. DePaolo
University of California, Berkeley

George M. Hornberger
The University of Virginia

ATTACHMENT C

Department of Energy
Washington, DC 20585

AUG 0 9 1996

AUG 1 3 1996

Dr. E. William Colglazier
National Academy of Sciences
2101 Constitution Avenue
Washington, DC 20418

Dear Dr. Colglazier:

National Academy of Sciences' Committee on Building and Environmental Management Science Program recently completed its initial assessment of the Departments's new basic research program in environmental management science. We are extremely pleased with the quality of this report, and we are grateful that the committee was able to complete it on such an accelerated schedule. The committee's recommendations have proven to be especially valuable to the Department in making award decisions in the FY96 proposal competition.

One of the report's principal recommendations was that the Department should postpone the release of the 1997 program announcement until it had time to identify and incorporate the lessons learned from the FY 1996 proposal competition, and to think more carefully, using the advice of the committee where appropriate, about how the Environmental Management Science Program should be structured and managed. As the Department begins work on the FY97 program announcement, it is most interested in receiving the committee's advice on the content of the announcement, particularly on the statement of research needs. The Department included a rather exhaustive list of research needs in the FY96 program announcement. We are especially interested in obtaining the committee's advice on how this list should be modified in order to better articulate these needs to researchers, many of whom have little or no knowledge of the weapons complex and the cleanup challenges.

The 1997 program announcement should be released by approximately October 31, 1996 in order to give researchers enough time to prepare and submit proposals. In order to meet this release date, the Department must finalize the announcement by October 11, 1996.

Thus, it would be most helpful if the committee could provide this
advise in a short report to the Department no later than
October 1, 1996.

I appreciate your consideration of this request.

Sincerely,

Carol J Henry

Carol J. Henry Ph.D., D.A.B.T.
Associate Deputy Assistant Secretary
  for Science and Risk Policy
Office of Science and Technology
Office of Environmental Management

cc: Kevin Crowley

ATTACHMENT D

Example Proposal Format

Project Abstract

Project Narrative

    Goals

    Scientific Significance of Project

    Relevance of Project to the EM Cleanup Mission

    Background

    Research Plan

        Preliminary Studies (if applicable)

        Literature Cited

        Research Design and Methodologies

        Collaborative Arrangements (if applicable)

Appendices

    Biographical Sketches

    Description of Facilities and Resources

    Budget

    Budget Explanation

    Current and Pending Support

ATTACHMENT E

Future Funding Scenarios for the
Environmental Management Science Program

The purpose of this attachment is to illustrate two scenarios for funding of the Environmental Management Science Program (EMSP) through FY 2002 by extrapolating, under two sets of assumptions, from the FY 1996 award results. The objective of these scenarios is to illustrate some consequences of the "mortgage" problem created by the "outyear" (i.e., post FY 1996) funding commitments made in the FY 1996 proposal competition.[1]

The *unconstrained funding scenario,* which is shown in Table E.1, was generated using the following set of assumptions:

• Funding of new awards for non-DOE performers (i.e., university, industry, and nonprofit performers) is continued at the FY 1996 level of $43 million for 3-year grants, and these awards are funded fully in the first year, as was the case for the FY 1996 proposal competition.
• The ratio of dollars committed each year to awards to non-DOE performers to the dollars committed each year to new awards to national lab performers remains constant at FY 1996 levels.[2]
    • Awards to national lab performers are paid in equal installments over 3 years.
    • Total annual funding for the EMSP is allowed to increase as necessary to satisfy the foregoing assumptions.

As shown in Table E.1, in order to maintain funding at FY 1996 levels, the total annual funding for the program would almost triple, to $131 million in FY 1999, before declining to a steady-state value of $112 million in FY 2000. This amount is roughly 225 percent of the current annual budget for the program.

The *constrained funding scenario,* which is shown in Table E.2, was generated using the following set of assumptions:

    • Total annual program funding is constrained to FY 1996 levels of $50 million.[3]

---

[1] In FY 1996, the DOE committed a total of $112 million to the EMSP. A total of $43 million was awarded to non-DOE performers, and these awards were funded fully in FY 1996. A total of about $4 million was provided to national lab performers in FY 1996. The remaining $63 million dollars in funding to national laboratory performers will be provided from FY 1997, FY 1998, and FY 1999 program funds as shown in Tables E.1 and E.2.

[2] In FY 1996, $43 million was awarded to non-DOE performers and $69 million was awarded to national lab performers. The ratio of dollars awarded is thus about 0.62.

[3] In FY 1996, $47 million of the $50 million in program funds were awarded to non-DOE and national laboratory performers. The remaining $3 million was used for other program-related purposes. To simplify the analysis, the committee assumes that all program funds are awarded to researchers in future years.

Dr. Carol Henry
October 8, 1996
page E-2

• As in the unconstrained funding scenario, the ratio of dollars committed each year to awards to non-DOE performers to the dollars committed to new awards to national laboratory performers remains constant at FY 1996 levels.
• As in the unconstrained funding scenario, awards to national laboratory performers are paid in equal installments over 3 years. The first installment is paid during the fiscal year in which the awards were made. The two remaining installments are paid in the two succeeding fiscal years. As shown by the scenario in Table E.1, for example, the $69 million awarded to national laboratories in FY 1997 would be paid in three equal installments of $23 million in FY 1997, $23 million in FY 1998, and $23 million in FY 1999.

This scenario illustrates the full effects of the mortgage when national laboratory performers receive funding one year at a time and non-DOE performers receive all of their funding up front. As shown in Table E.2, the mortgage from the FY 1996 award cycle creates a significant drain on program funds through FY 1999. Indeed, by FY 1999 only $10 million in new funds are available to non-DOE performers and $6 million in new funds are available to national laboratory performers, about a quarter of the funding available in FY 1996.[4]

The committee believes that the following conclusions can be inferred reasonably from the scenarios shown above: (1) Funding for the program will have to increase significantly in future years (e.g., as shown in Table E.1) in order to maintain current levels of program funding and a reasonable distribution of funding between non-DOE and national lab performers; or (2) both non-DOE and national lab performers will see a significant drop in funding for new or competetive renewal projects (e.g., Table E.2) if total annual funding for the program remains constant or decreases.

---

[4] As shown on Table E-2, an additional $12 million in funding commitments would be made to national laboratories in the FY 1999 program competition, but availability of these funds would depend on future congressional appropriations.

Dr. Carol Henry
October 8, 1996
page E-3

TABLE E.1 Hypothetical Funding for the EMSP when annual program funding is unconstrained.

Program	Funds Distributed During Fiscal Year (millions of dollars)						
Fiscal Year	1996	1997	1998	1999	2000	2001	2002
Non-DOE performers							
1996*	43	0	0	0	0	0	0
1997		43	0	0	0	0	0
1998			43	0	0	0	0
1999				43	0	0	0
2000					43	0	0
2001						43	0
2002							43
National lab performers							
1996*	4	23	23	19	0	0	0
1997		23	23	23	0	0	0
1998			23	23	23	0	0
1999				23	23	23	0
2000					23	23	23
2001						23	23
2002							23
TOTAL	47	89	112	131	112	112	112

* Results from the FY 1996 proposal competition.

Dr. Carol Henry
October 8, 1996
page E-4

TABLE E.2 Hypothetical Funding for the EMSP when annual program funding is constrained to $50 million.

| Program | Funds Distributed During Fiscal Year (millions of dollars) | | | | | | |
Fiscal Year	1996	1997	1998	1999	2000	2001	2002
Non-DOE performers							
1996*	43	0	0	0	0	0	0
1997		18	0	0	0	0	0
1998			12	0	0	0	0
1999				10	0	0	0
2000					25	0	0
2001						20	0
2002							17
National lab performers							
1996*	4	23	23	19	0	0	0
1997		9	9	9	0	0	0
1998			6	6	6	0	0
1999				6	6	6	0
2000					13	13	13
2001						11	11
2002							9
TOTAL	47	50	50	50	50	50	50

* Results from the FY 1996 proposal competition.

APPENDIX H

ACRONYMS

BEMR	*The 1996 Baseline Environmental Management Report*
DOD	U.S. Department of Defense
DOE	U.S. Department of Energy
DOE-EM (EM)	U.S. Department of Energy, Office of Environmental Management
DOE-ER (ER)	U.S. Department of Energy, Office of Energy Research
EM-50	U.S. Department of Energy, Office of Environmental Management, Office of Science and Technology
EMSP	U.S. Department of Energy, Environmental Management Science Program
EPA	U.S. Environmental Protection Agency
FACA	Federal Advisory Committee Act
FY	fiscal year (for the U.S. government, October 1 of a given year through September 30 of the following year)
GAO	U.S. General Accounting Office
GPO	U.S. Government Printing Office
GPRA	Government Performance and Results Act
HLW	high-level waste
H.R.	U.S. House of Representatives Bill
INEL	Idaho National Engineering Laboratory
NASA	National Aeronautics and Space Administration
NABIR	Natural and Accelerated Bioremediation Program
NAS	National Academy of Sciences
NBS	National Bureau of Standards
NIH	National Institutes of Health
NIST	National Institute of Standards and Technology
NRC	National Research Council
NSF	National Science Foundation

ONR	Office of Naval Research (DOD)
OSTP	Office of Science and Technology Policy
P.I.	principal investigator
PNL	Pacific Northwest Laboratory
R&D	research and development
TRU	transuranic
USGS	U.S. Geological Survey
WIPP	Waste Isolation Pilot Plant